# EURO Advanced Tutorials on Operational Research

**Series Editors**

The EURO Advanced Tutorials on Operational Research are a series of short books devoted to an advanced topic—a topic that is not treated in depth in available textbooks. The series covers comprehensively all aspects of Operations Research. The scope of a Tutorial is to provide an understanding of an advanced topic to young researchers, such as Ph.D. students or Post-docs, but also to senior researchers and practitioners. Tutorials may be used as textbooks in graduate courses. All titles in the series are peer-reviewed.

The series is SCOPUS-indexed.

Andrianos E. Tsekrekos •
Athanasios N. Yannacopoulos

# Variational Inequalities in Management Science and Finance

## Modelling, Analysis, Numerics and Applications

Springer

Andrianos E. Tsekrekos
Department of Accounting and Finance
Athens University of Economics
and Business
Athens, Greece

Athanasios N. Yannacopoulos
Department of Statistics
Athens University of Economics
and Business
Athens, Greece

ISSN 2364-687X ISSN 2364-6888 (electronic)
EURO Advanced Tutorials on Operational Research
ISBN 978-3-031-74118-0 ISBN 978-3-031-74119-7 (eBook)
https://doi.org/10.1007/978-3-031-74119-7



This Springer imprint is published by the registered company Springer Nature Switzerland AG
The registered company address is: Gewerbestrasse 11, 6330 Cham, Switzerland

*A.E.T: To Manolis and Thanasis*

*A.N.Y: To the magnificent four*

# Preface

*A giant leap for the authors, an insignificant step for humanity —paraphrasing Alessandro Gerosa, author of the Hipster Economy*

The aim of this EURO Advanced Tutorial on Operational Research is to provide readers with a concise and rigorous introduction to the theory, computation, and applications of variational inequalities (VIs), with emphasis on applications in management science and finance. It aims to bridge the gap between abstract mathematical treatments of the subject and simplistic, non-rigorous approaches often used in financial economics or managerial literature. Building on a number of concrete applications drawn from management science and finance, that will serve as motivating examples, the proposed tutorial will gradually develop the modelling of concrete problems in terms of variational inequalities in a general form, the foundation of an analytical theory both in finite and infinite dimensions, concerning their qualitative properties, and their rigorous numerical treatment, building on this analysis. After completing this course, the readers will have the necessary tools to embark on the modelling and analysis of concrete problems in management science and finance in terms of the powerful tool of VIs, and follow the current research literature on this fascinating and rich field.

The book consists of four chapters. The first one, entitled Modelling, introduces the concept of a variational inequality (VI), in terms of three motivating examples: equilibria in transportation networks, Nash equilibria in noncooperative games, and optimal stopping problems (related to the valuation of American and real options). We will return to these three examples during the remainder of the book, using them as testbeds or examples for illustrating the material introduced and developed as we proceed. The second chapter, entitled Analysis, develops the analytical framework for the study of crucial problems related to the well posedness (existence and uniqueness) as well as regularity of solutions for classes of VIs, in finite or infinite dimensional spaces. These techniques are illustrated with examples, related to the three motivating examples introduced in the first chapter. Chapter 3 focuses on the numerical treatment of VIs, building upon the analytic theory developed in Chap. 2. It offers an introduction to the rigorous numerical analysis of VIs, based around two classes of algorithms, proximal algorithms with operator splitting and duality based algorithms, providing a unified framework for them in terms of maximal monotone operator theory and rigorous results as to their convergence. Full and intuitive proofs of the main results in Chaps. 2 and 3 are provided, trying to guide the reader through the concepts and aiming to gradually build up the required skills for dealing with this fascinating subject. The required technical mathematical toolbox is provided as we go along, trying at each step to help the reader understand where and why this is needed. The final chapter is a short selection of more modern topics and extensions, selected from the current research literature, with its main aim not to offer the reader a full view of these topics (which would anyway be impossible in a short tutorial such as this one) but rather show to the reader how the concepts introduced in Chaps. 2 and 3 can be put to action and perform state-of-the-art research (or at least understand the latest issue of "*Annals of Improbable Research*" that just arrived at the library).

We tried to keep the presentation of the material as non-technical as possible, without losing rigour. Hopefully, we succeeded in providing a presentation of this material which leads the reader through the (necessary) technicalities gradually, but in a way that makes clearly understood where and why and to which extent these are required. The concepts and techniques introduced are illustrated through numerous examples in the text, and the main results are then applied to the treatment of the three main motivating examples introduced in the beginning, which span a wide range of the applicability of VIs in management science and finance. We hope that the output will help the target audience of the book, newcomers to the field of VIs and their many applications, realize their beauty and usefulness and provide them with a solid understanding of the foundations needed to delve deeper into it. The intended audience of the book is advanced M.Sc. and Ph.D. students, researchers in academia, or the industry wishing or needing to use VIs in their modelling work or anyone really that wants a peek into this fascinating universe. It can serve as a text in support for a short course but is also ideal for self-study. We hope you find it useful and enjoyable.

Athens, Greece
July 2024

Andrianos E. Tsekrekos
Athanasios N. Yannacopoulos

# Acknowledgements

We wish to thank the editors of the series "EURO Advanced Tutorials in Operational Research," Professor M. Grazia Speranza and Professor Kenneth Sorensen (and also the previous co-editor, Professor José Fernando Oliveira) for their kind invitation to publish this book in the series and for their help and support during the preparation of this book. We would also like to sincerely thank Drs Georgios Domazakis and Kyriakos Georgiou for critical reading of parts of the book and their constructive comments. ANY wishes to thank Professor Electra Petracou for her invaluable support and Professor Em. Nikos Yannacopoulos for advice on game theory and not only.

# Contents

# Chapter 1
# Modelling

*He who seeks for methods without having a definite problem in mind seeks for the most part in vain.*

*David Hilbert*

*Mathematicians are like Frenchmen: whatever you say to them they translate into their own language and forthwith it is something entirely different.*

*Johann Wolfgang von Goethe*

## 1.1 What and Why?

Our starting point will be the two fundamental questions: what and why?

### *1.1.1 What?*

What are variational inequalities and why do we need them in the modelling of decision-making processes in management science and finance? We start with a formal definition, which we will try to motivate through three worked modelling exercises inspired from the above fields.

**Definition 1.1 (Variational Inequality)** Given a convex closed subset $C \subset \mathbb{R}^d$ and a continuous function $F : C \to \mathbb{R}^d$, a variational inequality (VI) problem is stated as

$$\text{Find } x^* \in C \ : \ \langle F(x^*), x - x^* \rangle \geq 0, \ \forall x \in C, \tag{1.1}$$

where by $\langle \cdot, \cdot \rangle$ we denote the inner product in $\mathbb{R}^d$.

(continued)

 
A. E. Tsekrekos, A. N. Yannacopoulos, *Variational Inequalities in Management Science and Finance*, EURO Advanced Tutorials on Operational Research,
https://doi.org/10.1007/978-3-031-74119-7_1

**Definition 1.1** (continued)
**Notation:** Variational inequalities will be abbreviated to VIs. The VI (1.1) will be denoted by $VI(F, C)$ and the set of solutions by $SOL(F, C)$.

The problem makes perfect sense and finds important applications in the general case where $\langle \cdot, \cdot \rangle$ is no longer the standard inner product[1] but rather *any* inner product in $\mathbb{R}^d$. It also makes sense and finds important applications when $\mathbb{R}^d$ is replaced by $X$, a Hilbert space with inner product $\langle \cdot, \cdot \rangle$, in either finite or infinite dimensions. Both cases will be covered in this book trying to equilibrate as much as possible between modelling, analysis, numerics, and applications of VIs in management science and finance.

**Do Remember 1.2** *Let $X$ be a normed vector space, and let $C \subset X$.*

- *$C$ is closed if for any $(x_n)_{n \in \mathbb{N}} \subset C$, such that $x_n \to x$, it holds that $x \in C$.*
- *$C$ is convex if $\lambda x_1 + (1 - \lambda x_2) \in C$, $\forall x_1, x_2 \in C$, $\forall \lambda \in [0, 1]$.*

Normed vector spaces are large classes of sets that include $\mathbb{R}^d$ or Hilbert spaces.

Various objects that we have already encountered in various applications can be brought into the form of a VI such as (1.1).

**Example 1.3 (Equations)** Suppose that $C = E$, a linear subspace of $\mathbb{R}^d$. Then, choosing any $z \in E$ and setting $x = x^* \pm z$ in (1.1), we see that

$$\pm \langle F(x^*), z \rangle \geq 0, \quad \forall \, z \in E \iff \langle F(x^*), z \rangle = 0, \quad \forall \, z \in E.$$

If $E = \mathbb{R}^d$, this immediately implies that $F(x^*) = 0$, i.e., the VI reduces to a system of equations. This remains true in the case where $X$ is a general Hilbert space. $\lhd$

**Example 1.4 (Complementarity Conditions)** Suppose that $C = \mathbb{R}^d_+$, the positive cone of $\mathbb{R}^d$. Then, the VI (1.1) is equivalent to the complementarity problem

$$F(x^*) \geq 0, \quad x^* \geq 0, \quad \langle F(x^*), x^* \rangle = 0. \tag{1.2}$$

The conditions in (1.2) are called (non-linear) complementarity conditions.

---

[1] This can also be denoted as the dot product $\cdot$, $\langle x, y \rangle = x \cdot y$. For the most part of this book, however, we will employ the notation $\langle \cdot, \cdot \rangle$, to emphasize the general applicability of most of these results for more general cases.

Indeed, if $x^* \in \mathbb{R}^d_+$ satisfies (1.1), (a) we apply the inequality for $x = x^* + e_i \in \mathbb{R}^d_+$, $i = 1, \cdots, d$, where by $e_i$ we denote the vectors of the standard basis of $\mathbb{R}^d$ to obtain $F_i(x^*) \geq 0$ for all $i = 1, \cdots, d$, and therefore $F(x^*) \geq 0$, (b) for $x = 2x^* \in \mathbb{R}^d_+$ and (c) $x = 0 \in \mathbb{R}^d_+$ which yield $\pm\langle F(x^*), x^*\rangle \geq 0$, respectively, so that $\langle F(x^*), x^*\rangle = 0$. Conversely, if $x^* \in \mathbb{R}^d_+$ satisfies (1.2) then, for any $x \in \mathbb{R}^d_+$,

$$\langle F(x^*), x - x^*\rangle = \langle F(x^*), x\rangle - \langle F(x^*), x^*\rangle = \langle \underbrace{F(x^*)}_{\geq 0}, \underbrace{x}_{\geq 0}\rangle \geq 0,$$

which is (1.1). This remains true in the case where $X$ is a general Hilbert space that supports an order relation analogous to $\geq$. ◁

### 1.1.2 Why?

The general answer to the question "why" comes from the connection of certain types of VIs of the form (1.1) with optimization problems under constraints, by recognizing (1.1) as a condition characterizing the solution of an appropriate optimization problem. This can be done in various ways, and we mention a few of these.

**VIs and Optimization I: VIs as First Order Conditions for Constrained Minimization**

Although this connection may be restricted to specific cases of (1.1), we start our motivation from there, as it provides a very clear reason for the need to consider VIs.

**Theorem 1.5** *Let $C \subset \mathbb{R}^d$ be a closed convex set and $F : C \to \mathbb{R}^d$ be $F = D\Phi$, the gradient of the $C^1(\mathbb{R}^d)$ (continuously differentiable) function $\Phi : C \to \mathbb{R}$.*

*(i) If $x^* \in C$ is a solution to the constrained optimization problem*

$$\min_{x \in C} \Phi(x), \tag{1.3}$$

*then $x^*$ solves the VI*

$$\langle F(x^*), x - x^*\rangle \geq 0, \quad \forall x \in C, \quad \text{where } F(x) = D\Phi(x). \tag{1.4}$$

*(ii) If $\Phi$ is convex, then a solution of (1.4) is a solution of (1.3).*

***Proof***

(i) Consider a perturbation of the minimizer $x^* \in C$ of the form $x_\epsilon = x^* + \epsilon(x - x^*)$ for any $\epsilon > 0$ and any $x \in C$. Note that by the convexity of $C$, it follows that $x_\epsilon \in C$. Since $x^* \in C$ is a minimizer of problem (1.3), it holds that

$$\Phi(x_\epsilon) - \Phi(x^*) = \Phi(x^* + \epsilon(x - x^*)) - \Phi(x^*) \geq 0, \quad \forall \epsilon > 0, \quad x \in C. \tag{1.5}$$

Dividing both sides of (1.5) by $\epsilon > 0$ and passing to the limit as $\epsilon \to 0$ yield

$$\lim_{\epsilon \to 0^+} \frac{1}{\epsilon}(\Phi(x^* + \epsilon(x - x^*)) - \Phi(x^*)) \geq 0, \quad \forall x \in C, \tag{1.6}$$

where the existence of the limit is guaranteed by the differentiability of $\Phi$. Note that, by the same argument,

$$\lim_{\epsilon \to 0^+} \frac{1}{\epsilon}(\Phi(x^* + \epsilon(x - x^*)) - \Phi(x^*)) = \langle D\Phi(x^*), x - x^* \rangle,$$

so that (1.6) leads to

$$\langle D\Phi(x^*), x - x^* \rangle \geq 0, \quad \forall x \in C,$$

which is (1.4).

(ii) Recall that for a convex function

$$\Phi(x) - \Phi(z) \geq \langle D\Phi(z), x - z \rangle, \quad \forall\, x, z \in \mathbb{R}^d. \tag{1.7}$$

Apply this for $z = x^* \in C$, where $x^*$ solves (1.4) and any $z \in C$. Using (1.4) in (1.7), we get that $\Phi(x) - \Phi(x^*) \geq 0$ for every $x \in C$; hence, $x^*$ solves (1.3). □

**Do Remember 1.6** *A function* $f : \mathbb{R}^d \to \mathbb{R}$ *is:*

- *Convex if* $f(\lambda x + (1-\lambda)z) \leq \lambda f(x) + (1-\lambda) f(z), \ \forall x, z \in \mathbb{R}^d, \ \lambda \in [0, 1]$.
- $f \in C^1$ *: convex if and only if* $f(z) \geq f(x) + \langle Df(x), z - x \rangle$ *for all* $x, z \in \mathbb{R}^d$.
- $f \in C^1$ *: convex if and only if* $\langle Df(z) - Df(x), z - x \rangle \geq 0$ *for all* $x, z \in \mathbb{R}^d$.

The connection between variational inequalities and constrained optimization problems can be extended for the case where $X$ is a general Hilbert space (finite or infinite dimensional[2]). A prime example of infinite dimensional problems are optimal stopping problems, and the corresponding free boundary value problems that find important applications in American option pricing, or management decisions within the theory of real options (see Sect. 1.2.3).

**VIs and Optimization II: Merit or Gap Functions**

One may link VIs to minimization problems in other ways. An interesting way to achieve this leads to the concept of merit or gap functions, which is useful to both the analysis and the numerics of VIs.

We start by the observation that if $x^* \in C$ satisfies (1.1), then

$$\langle F(x^*), x - x^* \rangle \geq 0, \quad \forall x \in C \iff x^* \in \arg\min_{x \in C} \Phi(x),$$
$$\Phi(x) := \langle F(x^*), x - x^* \rangle, \tag{1.8}$$

i.e., the solution of the VI is characterized in terms of a constrained minimum of the function $x \mapsto \Phi(x)$ with $\Phi(x^*) = 0$. Even though this may seem as a bit of a swindle, since the function $\Phi$ depends on the (unknown) solution $x^*$ as a parameter, this simple observation may have important consequences.

We can get round this paradoxical situation as follows: Extend $\Phi$ by doubling the variables, consider instead the new function $L$ defined as

$$(x, y) \mapsto L(x, y) := \langle F(x), x - y \rangle, \tag{1.9}$$

and then define the function $x \mapsto \gamma_A$ by

$$\gamma_A(x) = \max_{y \in C} L(x, y) = \max_{y \in C} \langle F(x), x - y \rangle. \tag{1.10}$$

The function $\gamma_A$ is directly related to the solution of (1.1).

**Proposition 1.7** *The function $\gamma_A$ in* (1.10) *has the following properties:*

*(i)* $\gamma_A(x) \geq 0$ *for all* $x \in C$

*(ii)* $\gamma_A(x^*) = 0$ *if and only if* $x^* \in SOL(F, C)$,

*i.e.,* $VI(F, C)$ *(problem* (1.1)*) can be expressed as the minimization problem*

$$\min_{x \in C} \gamma_A(x). \tag{1.11}$$

[2] In the latter case $DF$ is the Gâteaux derivative (see, e.g., Kravvaritis and Yannacopoulos (2020)).

***Proof*** (i) follows from the fact that $x \in C$; hence by definition $\gamma_A(x) \geq L(x, x) = 0$. For (ii), if $\gamma(x^*) = 0$, then $\max_{y \in C} L(x^*, y) = 0$; hence $L(x^*, y) \leq 0$ for every $y \in C$ which implies that $x^* \in SOL(F, C)$. Conversely, if $x^* \in SOL(F, C)$, then $L(x^*, y) \leq 0$ for every $y \in C$, so $\max_{y \in C} L(x^*, y) \leq 0$, but since $L(x^*, x^*) = 0$ for $x^* \in C$, we conclude that $\gamma_A(x^*) = 0$. □

Moreover, by the very definition of $\gamma_A$, the minimization problem (1.11) can be written in the max-min formulation

$$\min_{x \in C} \gamma_A(x) = \min_{x \in C} \max_{y \in C} L(x, y) \stackrel{?}{=} \max_{y \in C} \min_{x \in C} L(x, y), \tag{1.12}$$

with the last equality subject to the validity of a minimax theorem for $L$ and $C$ (hence the question mark), and the resulting problem is called the dual problem. This essentially leads to the possibility of reformulating the VI in terms of a saddle point formulation.

This formulation is very interesting as it opens the road to connecting the study of VIs with minimax arguments and in terms of duality formulations, a theme that is useful from the point of view of both analysis and numerics. We shall return to this subject in the next two chapters.

What makes this formulation more interesting is the fact that the choice for the function $\gamma_A$ may not be unique!

**Definition 1.8 (Gap Function)** Given $VI(F, C)$, a function $\gamma$ satisfying properties (i) and (ii) of Proposition 1.7 is called a gap function for $VI(F, C)$.

Given a gap function $\gamma$ for $VI(F, C)$ allows for alternative analytic or numerical treatments for the VI. For example, one may use a minimization algorithm to minimize the gap function and thus obtain a solution of the VI.

Moreover, different gap functions have different properties and are possibly suitable for different usage. For example, the gap function introduced here, called the Auslender gap function,[3] is not smooth. However, there is a range of alternative gap functions, for example, the gap function proposed by Fukushima (see, e.g., Fukushima (1996) and the references therein)

$$\gamma_F(x) = \max_{y \in C} \left( \langle F(x), x - y \rangle - \frac{\alpha}{2} \|x - y\|^2 \right), \quad \alpha > 0,$$

which is a regularization of the gap function $\gamma_A$, which presents smoothness properties. Other choices are to develop gap function in terms of the dual problem

[3] Also known in the game theory community, see Larsson and Patriksson (1994) and the references therein.

either using a general Fenchel transform framework (see, e.g., Auchmuty (1989)) or in terms of Lagrangian duality for sets defined in terms $C = \{g_i(x) \leq 0,\ i = 1, \cdots, m\}$, where $g_i$ is convex (see, e.g., Auslender and Teboulle (2000)). For a general account of gap functions, see also Facchinei and Pang (2003) and Goh (2002). Gap and duality methods can also be extended for infinite dimensional problems. We will return to this subject in Chap. 3.

## 1.2 Applications in Management Science and Finance

### *1.2.1 Equilibrium in Transportation Networks*

Problems related to the flow of some "entity" on networks find many applications in management and finance. The notion of entity and that of network can be fairly general. For instance, the quantity can correspond to a physical good in a logistics network, energy (e.g., electricity or natural gas) in an energy transportation network such as an electricity grid, some utility (e.g., water) in a utility supply network, money or other securities in a financial network, or even to less tangible entities such as information or bandwidth in a communications or computer network (see, e.g., Nagurney (2013) and the references therein). Here we present a very schematic and general view of a transportation network with the main aim of defining the concept of equilibrium and its connection with VIs.

**Definition 1.9 (Transportation Network)** A transportation network consists of:

(i) A graph $G = (N, E)$ comprising
   - The set of nodes $N$ (transportation sites) and
   - The set of edges $E \subset N \times N$ (direct links between transportation sites)

(ii) A set $W \subset N \times N$ (origin-destination (O/D) points)
(iii) A demand structure $d \in \mathbb{R}^{|W|}$ for each O/D site and
(iv) A cost vector $c \in \mathbb{R}^{|E|}$ for using each link depending on the flow pattern of the network

**Notation 1.10** *For any set A, we will denote by $|A|$ its cardinality number.*

In typical applications, the structure of the graph $G$, the demand pattern $d$, and the cost function $c$ are given, and one has to determine a flow pattern of the entity, covering the required demand at each O/D point and satisfying certain equilibrium conditions or optimizing certain criteria (e.g., minimizing some cost function). Problems of this type have been extensively studied in the literature in various forms and varieties.

The above concise definition of a transportation network requires some clarifications and an example. We will start with the abstract general formulation and then present the example; depending on the reader's sophistication, the reader may switch from one to the other ad lib.

The graph $G$ is directed, i.e., if $e := (i, j) \in E$ is a direct link $i \to j$, this does not necessarily imply that $e' = (j, i) \in E$, i.e., a direct link $j \to i$ may not exist. Nodes may also be connected indirectly, in terms of paths. A path connecting $i$ with $j$ is a (acyclic) sequence of links $(i, i_1), (i_1, i_2), \cdots, (i_k, j)$, with each $(i_{k-1}, i_k) \in E$, which starts at $i$ and ends at $j$, thus corresponding to a route that connects $i$ to $j$ indirectly. There may be multiple routes between any two nodes $i$ and $j$; for each $w = (i, j) \in W$, we will define:

- $P_w$ the set of paths $p_w$ starting from the node $i$ and going to the node $j$ and
- $P = \bigcup_{w \in W} P_w$ the set of all paths on the graph

More compactly, a path $p \in P_w$ can be expressed as $p = \{e_1, \cdots, e_k\}$ with $e_1 = (i, \cdot) \in E$ and $e_k = (\cdot, j) \in E$. An origin-destination (O/D) point $w = (i, j) \in W$ will be called connected if $P_w \neq \emptyset$. In other words, if two nodes $i, j$ are not directly connected, then $(i, j) \notin E$, but if for $w = (i, j)$, $P_w \neq \emptyset$, then there can be transport from $i \to j$ in terms of one or more of the paths $p \in P_w$ that indirectly connect the two nodes.

The activity on the network is characterized by flow of the good $x_p$ on each path $p \in P$. Note that if $p \in P_w$ for some $w = (i, j) \in W$, then $x_p$ is to be understood as the total quantity of the good that travelled from $i$ to $j$ which can only be consumed at $j$ and not at any intermediate point in the path: That means a parcel that has been allocated for a path $p$ scheduled from $i \to j$ cannot be split and used to cover any needs arising at some intermediate location $k$ within the path. The total activity in the network is collected in a vector $x \in \mathbb{R}_+^{|P|}$, called the flow pattern, and containing information on the quantity transported between O/D sites through all possible routes.

To each $w = (i, j) \in W$, we can assign a demand $d_w$ for transportation of the good from $i \to j$, which can be accommodated through flow of this quantity $x_{p_w}$ along any path $p_w \in P_w$, in terms of the book keeping equation (the total demand may be split into different paths)

$$d_w = \sum_{p_w \in P_w} x_{p_w} = \sum_{p \in P} \delta_{w,p} x_p, \quad \forall\, w \in W, \quad \text{where } \delta_{w,p} = \begin{cases} 1 \text{ if } & p \in P_w \\ 0 & \text{otherwise.} \end{cases} \tag{1.13}$$

The full demand information on the network is collected in a vector $d \in \mathbb{R}_+^{|W|}$, called the demand pattern, and containing in its components information on the demand for any O/D point $w \in W$. Clearly (1.13) should hold for every component of this vector. In terms of the above notation we may express the demand covering constraints in compact form as

$$d = Bx, \quad \text{where } B = (\delta_{w,p} \ : \ w \in W, p \in P) \in \mathbb{R}^{|W|\times|P|}. \tag{1.14}$$

A flow will be called feasible if $x$ satisfies (1.14), i.e., if

$$x \in C := \{x \geq 0 \ : \ d = Bx\} \subset \mathbb{R}_+^{|P|\times 1} = \mathbb{R}_+^{|P|}. \tag{1.15}$$

The flows induce a load $y_e$ on each link $e \in E$ that can be expressed in terms of

$$y_e = \sum_{p\in P} \delta_{e,p} x_p, \quad \forall\, e \in E, \quad \text{where } \delta_{e,p} = \begin{cases} 1 \text{ if } & e \in P \\ 0 \text{ otherwise.} \end{cases} \tag{1.16}$$

The loads on each link can be collected in a vector $y = (y_e \ : \ e \in E) \in \mathbb{R}_+^{|E|\times 1} = \mathbb{R}_+^{|E|}$, and we may express

$$y = Ax, \quad \text{where } A = (\delta_{ep} \ : \ e \in E, \ p \in P) \in \mathbb{R}^{|E|\times|P|}. \tag{1.17}$$

The matrix $A$ is an appropriate incidence matrix for the network. The vector $y$ is important as in many problems, e.g., in communication networks, etc., there may be important capacity constraints.

Clearly, there are transportation costs, which may vary from path to path and may depend on the flow pattern. This cost has to be broken down into the various links in $E$, comprising any chosen path $p$. It is a plausible assumption that the cost per link $e$ may depend on the load imposed on each individual link $y_e$ and perhaps also on the loads $y_{e'}$ on the other links $e'$ in a nonlocal way. In any case, the cost depends on $x$ through the relation $y = Ax$. We must then define for each $e \in E$ the cost function $c_e$ and the vector $c = (c_e \ : \ e \in E) \in \mathbb{R}_+^{|E|}$ collecting all the local transportation cost structures and characterizing the network. This is to be understood as a function of the flow pattern

$$x \mapsto c(x) = (c_e(y) \ : \ e \in E) = (c_e(Ax) \ : \ e \in E) \in \mathbb{R}_+^{|E|}.$$

The total cost $K_{p_w}$ of a path $p_w \in P_w$ can then be computed in terms of the sum of costs per link as $K_{p_w} = \sum_{e\in p_w} c_e = \sum_{e\in E} \delta_{ep_w} c_e$, so that upon defining the total cost vector $K = (K_p, \ p \in \bigcup_{w\in W} P_w) \in \mathbb{R}^{|P|\times 1}$, which collects the total cost over all possible paths, joining any $w \in W$, we get in compact matrix notation

$$K = A^T c(x) \overset{(1.17)}{=} A^T c(Ax). \tag{1.18}$$

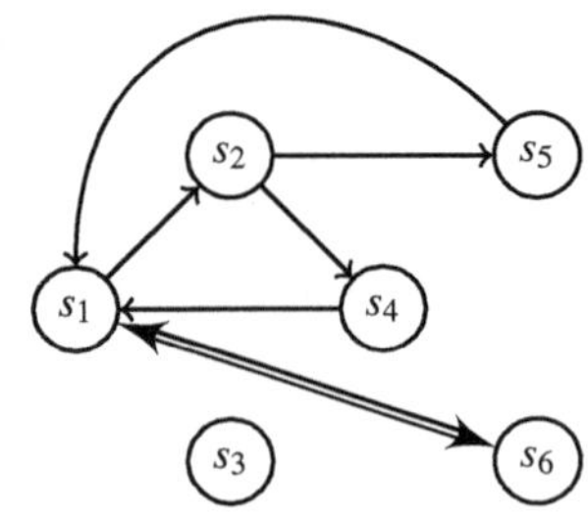


**Fig. 1.1** Example of a transportation network

**Example 1.11** As an example consider the transportation network shown in Fig. 1.1. In this example the set of nodes $N$, edges $E$, and O/D sites $W$ is

$$N = \{s_1, s_2, s_3, s_4, s_5, s_6\} = \{1, 2, 3, 4, 5, 6\},$$

$$E = \{(1, 2), (1, 6), (2, 4), (2, 5), (4, 1), (5, 1), (6, 1)\} = \{e_1, e_2, e_3, e_4, e_5, e_6, e_7\},$$

$$W = \{(2, 1), (1, 2), (5, 6), (5, 3)\} = \{w_1, w_2, w_3, w_4\},$$

$$|N| = 6, \ |E| = 7, \ |W| = 4.$$

There is no direct link $2 \to 1$ since $(2, 1) \notin E$, but there is an indirect link between them in terms of the path $p = \{(2, 4), (4, 1)\}$ or the path $p' = \{(2, 5), (5, 1)\}$. In fact if we are interested in transport between sites 2 and 1, i.e., $w = (2, 1) \in W$, then $P_w = \{p, p'\}$, is the set of all paths that may link $2 \to 1$. On the other hand, if $w' = (5, 3) \in W$, then $P_{w'} = \emptyset$ as there are no paths that may connect $5 \to 3$.

To proceed we need to adopt an enumeration of the paths.

$$w_1 = (2, 1), \ P_{w_1} = \{p_1, p_2\}, \ p_1 = \{(2, 5), (5, 1)\} = e_4 e_6,$$

$$p_2 = \{(2, 4), (4, 1)\} = e_3 e_5,$$

$$w_2 = (1, 2), \ P_{w_2} = \{p_3\}, \ p_3 = \{(1, 2)\} = e_1,$$

$$w_3 = (5, 6), \ P_{w_3} = \{p_4, p_5\}, \ p_4 = \{(5, 1), (1, 6)\} = e_6 e_2,$$

$$p_5 = \{(5, 1), (1, 2), (2, 4), (4, 1), (1, 6)\} = e_6 e_1 e_3 e_5 e_2,$$

$$w_4 = (5, 3), \ P_{w_3} = \emptyset,$$

$$P = \bigcup_{w \in W} P_w = P_{w_1} \cup P_{w_2} \cup P_{w_3} \cup P_{w_4} = \{p_1, p_2, p_3, p_4, p_5\}, \ \ |P| = 5.$$

For this simple setting we may clarify the $\delta$ notation. Using the above numbering $\delta_{w_1 p_1} = 1$ since $p_1 \in P_{w_1}$, whereas $\delta_{w_1, p_3} = 0$ since $p_3 \notin P_{w_1}$. Similarly $\delta_{e_3 p_1} = 0$ since $e_3 \notin p_1$, whereas $\delta_{e_3 p_2} = 1$ since $e_3 \in p_2$. All this information can be collected in the matrices

$$B = (\delta_{wp}) = \begin{pmatrix} 1\,1\,0\,0\,0 \\ 0\,0\,1\,0\,0 \\ 0\,0\,0\,1\,1 \\ 0\,0\,0\,0\,0 \end{pmatrix}, \qquad A = (\delta_{ep}) = \begin{pmatrix} 0\,0\,1\,0\,1 \\ 0\,0\,0\,1\,1 \\ 0\,1\,0\,0\,1 \\ 1\,0\,0\,0\,0 \\ 0\,1\,0\,0\,1 \\ 1\,0\,0\,1\,1 \\ 0\,0\,0\,0\,0 \end{pmatrix}.$$

The possible paths for transportation to the O/D sites are collected in the set $P$. Therefore the activity on this transportation network is fully described in terms of the flow vector $x = (x_1, x_2, x_3, x_4, x_5) \in \mathbb{R}^{|P|}$ which collects all the quantities transported through all available paths $p_1, p_2, p_3, p_4, p_5 \in P$. This has to be chosen so as to match the demand vector $d = (d_1, d_2, d_3, d_4) \in \mathbb{R}^{|W|}$, which contains information of the demand for each O/D site. Of course, out of these, $x_1, x_2$ cater for $w_1 = (2, 1)$ and can be considered as alternative or complementary ways of transporting the required quantity $d_{w_1}$ that is to be transported from $2 \to 1$. $x_3$ caters for $w_2 = (1, 2)$ and in this example is the only way that the demand $d_{w_2}$ is to be catered for. Finally there is no connectivity available for $w_3$, hence the zeros in the relevant rows of the matrix. Clearly, if $d_4 := d_{w_4} \neq 0$, then there is no way that this demand can be met in this network, so no flow pattern can be feasible for such a demand vector.

The loads are to be accounted for on the direct links, i.e., on the elements of $E$. For this example these are collected in a vector $y = (y_1, y_2, y_3, y_4, y_5, y_6, y_7) \in \mathbb{R}^{|E|}$, whose components are the total load created by the flow for each link. The load $y_i$ on link $e_i$ is created by the sum of all the flows on the paths that contain this link. For example, link $e_6$ participates in the paths $p_1$, $p_4$, and $p_5$. Hence, the total load on it is $x_1 + x_4 + x_5$, i.e., it depends on the total quantity that has passed from this link. The above argument easily leads to (1.17).

We finally consider the issue of cost. Each link in $E$ has a certain cost if used, which contributes to the total cost of a path. Certain paths may be cheaper than others, if they include cheaper links. All costs are collected in a vector $c = (c_1, \cdots, c_7) \in \mathbb{R}_+^{|E|}$. The cost of using each link may depend on the total load imposed to it by a flow pattern $x$, i.e., on the vector $y$. For example, given a flow pattern $x \in \mathbb{R}_+^{|P|}$, the cost of link $e_6$ must depend on $y_6 = x_1 + x_4 + x_5$, but it may also depend on the load of nearby patterns;[4] hence we will in general assume $c_6 = c_6(y)$ and not simply on $y_6$. As a simple example let us take here a linear cost function $c(y) = My$, where $M \in \mathbb{R}^{|E|\times|E|}$ is a matrix that provides the effect that the load of nearby links $e_j$, for $j \neq i$, has on the cost of a link $e_i$. If $M$ is a diagonal matrix, then the cost for link $e_i$ depends only on its load $y_i$. For this

[4] As, for example, in traffic flow, if a certain road is closed, other roads can be overloaded.

linear cost model, the cost function can be expressed in terms of the flow pattern $x$ as $c(x) = MAx$, and then the cost function of paths becomes $K(x) = A^T MAx$.

With that information we may calculate the cost of using each path in $P$. For example, using the path $p_5$ involves a total cost $K_5 = c_1 + c_2 + c_3 + c_6$ since this path uses the links $e_1, e_2, e_3, e_6$. On the other hand, using path $p_4$ involves a total cost $K_4 = c_2 + c_6$ since this path uses the links $e_2, e_6$. Since both paths cater for the same O/D sites (5, 6), when trying to cater for the demand $d_3$, we must regulate $x_4$ and $x_5$ so that $x_4 + x_5 = d_3$, but the cost $K_5, K_6$ that it is generated by this flow pattern is somehow minimal. ◁

We now consider the problem of, given the transportation network $(G, d, c)$, choosing the flow pattern $x$ in such a way that a certain optimality condition, e.g., in terms of cost minimization, is met. There are various types of equilibria that can be considered in transportation networks; here we focus on a particular type of equilibrium, called Wardrop equilibrium, which is some kind of user-defined equilibrium. Quoting Wardrop himself (Wardrop 1952),

> The journey times on all the routes actually used are equal, and less than those which would be experienced by a single vehicle on any unused route.

where in the above we can substitute journey times with some notion of cost.

Following Correa and Stier-Moses (2011), De Palma et al. (1998) we provide the following definition:

**Definition 1.12 (Wardrop Equilibrium)** A feasible flow allocation $x^* \in \mathbb{R}_+^{|P|}$ is a Wardrop equilibrium allocation if for each $w \in W$, there exists $\lambda_w$ such that

$$K_{p_w}(x^*) = \begin{cases} \lambda_w & \text{if } x^*_{p_w} > 0, \\ \geq \lambda_w & \text{if } x^*_{p_w} = 0, \end{cases} \quad \forall\ p \in P_w, \tag{1.19}$$

or equivalently,

$$K_{p_w}(x^*) \geq \lambda_w, \quad x_{p_w}(K_{p_w}(x^*) - \lambda_w) = 0, \ \ \forall\ p_w \in P_w,\ \ w \in W. \tag{1.20}$$

In compact form this is equivalent to the existence of a flow allocation $x^* \in C \subset \mathbb{R}_+^{|P|}$ and an equilibrium vector $\lambda^* \in \mathbb{R}^{|W|}$ such that

$$K(x^*) \geq B^T \lambda^*, \quad \langle x^*, K(x^*) - B^T \lambda^* \rangle = 0, \tag{1.21}$$

where $\langle \cdot, \cdot \rangle$ is the standard inner product in $\mathbb{R}^{|P|}$.[5]

[5] To go from (1.20) to (1.21), simply add all the equalities over all $p_w \in P_w$ and then over all $w \in W$. For the converse since all the components of the vectors $x^*$ and $K(x^*) - B^T \lambda^*$ are non-negative, the fact that the inner product vanishes implies that the products componentwise vanish as well, which leads to (1.20).

In the above $\lambda_w = \min_{p \in P_w} K_p$ is considered as the minimum cost that can accommodate a given demand pattern. This equilibrium condition states that if the cost of transporting the quantity along a path $p_w$ exceeds $\lambda_w$, then this path is not chosen, i.e., $x^*_{p_w} = 0$, whereas if it is equal to $\lambda_w$, it is adopted as a route and $x^*_{p_w} > 0$. According to this notion of equilibrium, all routes $p_w$ facilitating point $w \in W$ are expected to display the same minimal cost $\lambda_w$, with any route $p_w$ displaying higher cost being avoided.

This equilibrium allocation can be expressed as the solution of a variational inequality as follows:

**Proposition 1.13** *$x^* \in C$ is a Wardrop equilibrium if and only if it satisfies the variational inequality*

$$\langle K(x^*), x - x^* \rangle \geq 0, \quad \forall \, x \in C, \tag{1.22}$$

*where $C$ is the convex set defined in* (1.15) *and $K$ as in* (1.18).

*If the cost function $K : \mathbb{R}^{|P|} \to \mathbb{R}^{|P|}$ is continuous and monotone, i.e., $\langle K(x) - K(z), x - z \rangle \geq 0$ for all $x, z \in \mathbb{R}^{|P|}$, then* (1.22) *is equivalent to*

$$\langle K(x), x - x^* \rangle \geq 0, \quad \forall \, x \in C.$$

***Proof*** Let $x^* \in \mathbb{R}^{|P|}_+$ be a Wardrop equilibrium, and consider any feasible $x \in C$, i.e., satisfying

$$d = Bx = Bx^*. \tag{1.23}$$

Then, we have that

$$\begin{aligned} \langle K(x^*), x \rangle \overset{(1.21)}{\geq} \langle B^T \lambda, x \rangle = \langle \lambda, Bx \rangle \overset{(1.23)}{=} \langle \lambda, Bx^* \rangle \\ = \langle B^T \lambda, x^* \rangle \overset{(1.21)}{=} \langle K(x^*), x^* \rangle. \end{aligned}$$

For the converse, observe that if $x^* \in C$ solves (1.22), then $x^*$ solves the optimization problem

$$\min_x \langle K(x^*), x \rangle, \quad d = Bx, \quad x \geq 0. \tag{1.24}$$

Using the Lagrangian

$$L(x, \lambda, \mu) = L(x, \nu) = \langle K(x^*), x \rangle + \langle \lambda, d - Bx \rangle - \langle \mu, x \rangle,$$

we can express $x^*$ in terms of the saddle point $(x^*, \nu^*) := (x^*, \lambda^*, \mu^*) \in \mathbb{R}^{|P|} \times \mathbb{R}^{|W|} \times \mathbb{R}_+$ of $L$. The first order conditions for the minimization of $L$ with respect to $x$ yield

$$K(x^*) = B^T \lambda^* + \mu^*, \quad \mu^* \geq 0 \implies K(x^*) \geq B^T \lambda^*. \tag{1.25}$$

Moreover, the Kuhn-Tucker complementarity conditions imply that $\langle \mu^*, x^* \rangle = 0$, so taking the inner product with $x^*$ in (1.25) and using the complementarity condition we obtain $\langle x^*, K(x^*) - B^T \lambda^* \rangle = 0$.

The second assertion is a general result which is valid for monotone VIs, even outside the realm of optimal transportation networks, which is called Minty's reformulation or simply Minty's trick, that will be proved in its general form in Proposition 2.25. □

The reformulation of the Wardrop equilibrium in terms of a VI is important, as it allows us to use the general and well-studied theory and powerful numerical techniques developed for VIs (to be presented in Chaps. 2 and 3, respectively) for the study of Wardrop equilibria, obtain results concerning their properties, and design numerical schemes for their computation.

**Example 1.14** If the cost function $y \mapsto c(y)$ is linear, say $c(y) = My$ for a suitable matrix $M \in \mathbb{R}^{|E|\times|E|}$, then in terms of the flow pattern $x$ this function becomes $x \mapsto K(x) = A^T M A x =: K_0 x$. In this case the monotonicity of $K$ reduces to positivity of the matrix $K_0$. Then, the VI leading to the determination of Wardrop equilibrium reduces to

$$x^* \in C \;:\; \langle K_0 x^*, x - x^* \rangle, \;\; \forall\, x \in C = \{x \in \mathbb{R}_+^{|P|} \;:\; Bx - d = 0\}.$$

Returning to Example 1.11, if $M = diag(c_1, \cdots, c_7)$ for $c_i \geq 0$, $i = 1, \cdots, 7$, then

$$K_0 = \begin{pmatrix} c_4 + c_6 & 0 & 0 & c_6 & c_6 \\ 0 & c_3 + c_5 & 0 & 0 & c_3 + c_5 \\ 0 & 0 & c_1 & 0 & 0 \\ c_6 & 0 & 0 & c_2 + c_6 & c_2 + c_6 \\ c_6 & c_3 + c_5 & c_1 & c_2 + c_6 & c_1 + c_2 + c_3 + c_5 + c_6 \end{pmatrix},$$

leading to the corresponding VI. ◁

The topic of transportation networks is a very rich one, and its connection with VIs is widely studied; there exist many interesting extensions of the fundamental model introduced here. We refer the interested reader to, e.g., Nagurney (2013) and the references therein for a more detailed account.

## *1.2.2 Non-cooperative Games and Nash Equilibria*

The theory of games is an important tool in decision-making theory with many important applications in management science and finance (see, e.g., Mazalov (2014), Petracou and Yannacopoulos (2013), and the references therein). In contrast to standard optimization problems where a single decision-maker (DM) chooses a best strategy so as to minimize/maximize a certain cost/utility functional, in game theory we have a multitude of DMs, so that each DM, $i$, has to make a decision in the presence of the others, $j \neq i$, who impose constraints as to what decisions can be chosen by $i$ and/or the resulting pay-off.

**Definition 1.15 (Non-cooperative Game)** A non-cooperative game in normal form is given by a set of $m$ players, each characterized by a strategy set $K_i$ and a utility function $U_i : \prod_{j=1}^m K_j \to \mathbb{R}$, that the player wishes to maximize by choosing a strategy $x_i \in K_i$, constrained by the choices of all the other players $x_j \in K_j, j \neq i$.

**Notation 1.16** *We will use the notation* $\prod_{j=1}^m K_j := K_1 \times \cdots \times K_m$*, denote by* $x_{-i} := (x_1, \cdots, x_{i-1}, x_{i+1}, \cdots x_m) \in \prod_{j\neq i} K_j$ *the strategy vector of all other players but* $i$*, and express the utility function* $U_i$ *as* $U_i(x_i; x_{-i})$ *to emphasize and separate its dependence on the actions* $x_i$ *of player* $i$ *and the actions* $x_{-i}$ *of all the other players.*

Each player $i$ has only control over the strategy vector $x_i$ and often is unaware of the actions $x_{-i}$ of all the other players and may only postulate about them. Hence, given the actions $x_{-i}$ of all the other players, player $i$ can choose the strategy $x_i$ so as to maximize the utility $U_i(x_i; x_{-i})$, i.e., the best possible strategy for player $i$ would be

$$x_i^* \in \arg\max_{x_i \in K_i} U_i(x_i; x_{-i}). \tag{1.26}$$

As a mode of behavior we may consider that all players $j \in \{1, \cdots, m\}$ will act in the same way, taking into account that they have to decide given the actions $x_{-j}$ of all the other players, that of course decide according to the same rule as in (1.26). This concept leads to the celebrated Nash equilibrium (Nash 1951).

**Definition 1.17 (Nash Equilibrium)** A strategy vector $x^* = (x_1^*, \cdots, x_m^*) \in \prod_{j=1}^m K_j$ is called a Nash equilibrium if

$$x_i^* \in \arg\max_{x_i \in K_i} U_i(x_i; x_{-i}^*), \quad \forall\, i \in \{1, \cdots, m\}, \tag{1.27}$$

or equivalently

$$U_i(x_i^*; x_{-i}^*) \geq U_i(x_i; x_{-i}^*), \quad \forall\, x_i \in K_i, \ \forall\, i \in \{1, \cdots, m\}. \tag{1.28}$$

The strategy may be a pure strategy or a mixed strategy, understood as a probability distribution on strategy space. All these can be covered upon the choice of strategy space. Here, for simplicity, we restrict to the case where $K_i \subset \mathbb{R}^{d_i}$ for some finite $d_i$.

In words, a Nash equilibrium is a strategy from which none of the players wishes to deviate. Moreover, there is no guarantee that this equilibrium exists and is unique or that it is a desirable equilibrium; in fact, there are examples of games such as, for instance, the prisoner's dilemma game that a Nash equilibrium may be harmful to all players involved. There is also no guarantee that a Nash equilibrium satisfies other normative requirements, e.g., Pareto optimality. Quoting Ivar Ekeland,

> An equilibrium is a situation where each agent's actions turn out to be the best reply to everyone else's.[6]

to continue that

> An equilibrium is not always an optimum. It might not even be good. This may be the most important discovery of game theory.[7]

**Example 1.18 (Cournot Game)** This is a classic example, dating before the term game theory was introduced (Cournot 1838). Assume $m$ firms producing the same product, each with a strategy $x_i \in K_i = \mathbb{R}_+$, related to the quantity produced by firm $i$. The production of all firms targets the same market; hence the total quantity $x = \sum_{j=1}^m x_i = x_i + \sum_{j\neq i} x_j$ is released in the market. The price of the product is set in terms of a partial equilibrium approach; hence it depends on $x$. Assume a price function $x \mapsto P(x)$, for each unit of product. The utility of each firm $i$ can be chosen as $U_i(x_i; x_{-i}) = P(x)x_i - c_i x_i$, where $P(x) = P(x_i + \sum_{j\neq i} x_j) = P(x_i; x_{-i})$ and $c_i$ is the cost per unit of production for firm $i$. Note that $U_i(x_i; x_{-i})$ corresponds to the net profit of firm $i$ subject to its decision $x_i$ and the decisions $x_{-i}$ of all the other firms. Each firm $i$ solves the problem

$$\max_{x_i} \Big( P(x_i; x_{-i})x_i - c_i x_i \Big), \qquad i = 1, \cdots m,$$

where clearly each maximizer $x_i^*$ depends on the choice of the other firms $x_{-i}^*$.

As a fully worked example assume a linear price function $P(x) = a - \beta x$, for which case the first order conditions for the Nash equilibrium reduce to the linear system

$$\alpha - 2\beta x_i - \beta \sum_{j\neq i} x_j - c_i = 0, \;\; x_i \geq 0, \;\; i = 1, \cdots, m.$$

Note that the nature of the problem dictates to restrict only to solutions in the positive cone. ◁

---

[6] See Ekeland (2006), p. 140.

[7] See Ekeland (2006), p. 141.

Under certain conditions a Nash equilibrium can be expressed in terms of the solution of a suitable VI.

**Proposition 1.19** *Assume that the utility functions $U_i$ are continuously differentiable and that the functions $x_i \mapsto U_i(x_i; x_{-i})$ are concave. $x^* \in C$ is a Nash equilibrium if and only if it satisfies the VI*

$$\langle F(x^*), x - x^* \rangle_{\mathbb{R}^d} \geq 0, \ \ \forall \, x \in C, \tag{1.29}$$

*where*

$$F(x) = (-D_{x_1} U_1, \cdots, -D_{x_m} U_m) : \mathbb{R}^d \to \mathbb{R}^d, \ \ d = \sum_{i=1}^m d_i, \ \ C = \prod_{j=1}^m c_j.$$

***Proof*** Suppose $x^*$ is a Nash equilibrium. Then, each component $x_i^*$ maximizes $U_i(x_i; x_{-i}^*)$ or equivalently minimizes $-U_i(x_i; x_{-i}^*)$. By the differentiability of $U_i$, this is equivalent to the VI

$$\langle -D_{x_i} U_i(x_i^*; x_{-i}^*), x_i - x_i^* \rangle_{\mathbb{R}^{d_i}} \geq 0, \ \ \forall \, x_i \in K_i. \tag{1.30}$$

Adding over all players $i \in \{1, \cdots, m\}$, we obtain (1.29).

Conversely, let $x^* \in C$ satisfy (1.29). For any $i \in \{1, \cdots, m\}$ use as test element in (1.29), $x = (x_1^*, \cdots, x_{i-1}^*, x_i, x_{i+1}^*, \cdots, x_m^*) = (x_i; x_{-i}^*)$ for arbitrary $x_i \in C$. This yields (1.30) for every $i \in \{1, \cdots, m\}$ which, if $x_i \mapsto U_i(x_i; x_{-i})$ are concave (i.e., $x_i \mapsto -U_i(x_i; x_{-i})$ convex), implies (1.27) (see Do Remember 1.6). □

**Remark 1.20** The concavity assumption for $U_i$ can be weakened by assuming that $-U_i$ are pseudoconvex.[8] ⊕

**Example 1.21** The Nash equilibrium for the Cournot game of Example 1.18 is equivalent to the solution of the VI (1.29) for $C = \mathbb{R}_+^m$ and

$$F(x) = (-D_{x_1}(P(x)x_1) - c_1, \cdots, -D_{x_m}(P(x)x_m) - c_m).$$

In the particular case of linear demand function, the mapping $F$ is linear and of the form $F(x) = P_0 x + c$, where $P_0$ is the matrix

$$P_0 = \beta \begin{pmatrix} 2 & 1 & \cdots & 1 \\ 1 & 2 & \cdots & 1 \\ \vdots & \vdots & \ddots & \vdots \\ 1 & 1 & \cdots & 2 \end{pmatrix},$$

and $c = (c_1 - \alpha, \cdots, c_m - \alpha)$. ◁

[8] Recall that $f : \mathbb{R}^d \to \mathbb{R}$ is pseudoconvex if $\langle Df(x), z - x \rangle \geq 0$ implies $f(z) \geq f(x)$, for all $x, z \in \mathbb{R}^d$.

A Nash equilibrium is also related to a fixed point of a map. This map is constructed from the possible maximizers of each utility function, given all the other strategies. We define each player's $i$ best reply correspondence as

$$BR_i(x_{-i}) = \arg\max_{x_i \in K_i} U_i(x_i; x_{-i}).$$

In fact $BR_i(x_{-i})$ is a set (a subset of $K_i$), the set of maximizers of the function $x_i \mapsto U_i(x_i; x_{-i})$. This set depends on the strategies $x_{-i}$ adopted by the other players. The mapping $x = (x_1, \cdots, x_m) \mapsto \Phi_i(x) := BR_i(x_{-i})$ maps a point $x \in K_1 \times \cdots \times K_m$ to a subset of $K_i$; hence it is a point to set mapping, a multi-valued map. To emphasize this we will use the notation $BR_i : K_1 \times \cdots \times K_m \to 2^{K_i}$. If the function $x_i \mapsto U_i(x_i; x_{-i})$ has a unique maximizer for every $x_{-i}$, then the set $BR_i(x_{-i})$ is a singleton, and the mapping $BR_i$ is single-valued. In terms of the mapping $\Phi_i$, we can see that $x_i^*$ is a best reply of player $i$, given the strategy profile $x_{-i}$ of the other players, if

$$x_i^* \in \Phi_i(x_1, \cdots, x_{i-1}, x_i^*, x_{i+1}, \cdots, x_m) = BR_i(x_{-i}).$$

A Nash equilibrium $x^*$ satisfies

$$x_i^* \in \Phi_i(x_1^*, \cdots, x_{i-1}^*, x_i^*, x_{i+1}^*, \cdots, x_m^*) = BR_i(x_{-i}^*), \quad \forall\, i \in \{1, \cdots, m\};$$

hence upon defining the map $\Phi : K_1 \times \cdots \times K_m \to 2^{K_1} \times \cdots \times 2^{K_m}$ in terms of

$$\begin{aligned} x = (x_1, \cdots, x_m) \mapsto \Phi(x) &= (\Phi_1(x), \cdots, \Phi_m(x)) \\ &= (BR_1(x_{-1}), \cdots, BR_m(x_{-m})), \end{aligned}$$

we easily see that a Nash equilibrium $x^*$ is a fixed point of $\Phi$, i.e., a point satisfying the inclusion $x^* \in \Phi(x^*)$. This connection is important; in fact, as we shall see in Chaps. 2 and 3, there are deep connections between variational inequalities and fixed point problems, and in view of Proposition 1.19 this entails also a connection between fixed point theory and Nash equilibria.

**Nash Quasi-Equilibria and Quasi Variational Inequalities (QVIs)**

In the above analysis it is assumed that the strategy set $K_i$ for each player $i$ is fixed and constant, i.e., unaffected by the choice of strategy $x_{-i} \in \prod_{j \neq i} K_j$ of the remaining players. For certain games this may not be the case, and the strategy set of player $i$ may depend on $x_{-i}$, i.e., $K_i = K_i(x_{-i})$. This situation may appear in games arising in the management of common resources (e.g., a fishery, bandwidth in a communication network, etc.) or in games where there is a natural leader and follower structure. The concept of Nash equilibrium can be extended for such problems.

**Definition 1.22 (Nash Quasi-Equilibrium)** Let $K_i = K_i(x_{-i})$ be the strategy set of player $i$, given the strategy $x_{-i} \in \prod_{j \neq i} K_j$ adopted by the other players.

A strategy vector $x^* = (x_1^*, \cdots, x_m^*) \in \prod_{j=1}^m K_j$ is a Nash quasi-equilibrium if

$$x_i^* \in \arg \max_{x_i^* \in K_i(x_{-i}^*)} U_i(x_i^*; x_{-i}^*), \quad \forall\, i \in \{1, \cdots, m\}, \tag{1.31}$$

or equivalently,

$$U_i(x_i^*; x_{-i}^*) \geq U_i(x_i; x_{-i}^*), \quad \forall\, x_i \in K_i(x_{-i}^*), \quad \forall\, i \in \{1, \cdots, m\}. \tag{1.32}$$

**Example 1.23 (Harvesting a Non-renewable Resource)** Assume that we have $m$ players harvesting a non-renewable resource (e.g., an oil field), whose total quantity is $Q$. Each player harvests $x_i \in \mathbb{R}_+$ and derives a profit $P_i(x_i)$ out of it. On the other hand, harvesting impoverishes the resource, and this causes a disutility (loss) to each agent, of the form $L_i(Q - \sum_{j=1}^m x_j)$, where $Q - \sum_{j=1}^m x_j$ is the remaining stock of the resource and $z \mapsto L_i(z)$ is a decreasing function (i.e., low stock for the resource inflicts distress to the players). Each agent $i$ strives to maximize profit minus disutility, i.e., chooses $x_i$ so as to maximize $U_i(x_i; x_{-i}) := \Pi_i(x_i) - L_i(Q - x_i - \sum_{j \neq i}^m x_j)$. However here both the utility functions $U_i$ and the strategy sets $K_i$ depend on the actions of the players. The available choice for $i$ is $0 \leq x_i \leq Q - \sum_{j \neq i} x_j$, i.e., $x_i \in K_i(x_{-i})$, where $K_i(x_{-i}) = [0, Q - \sum_{j \neq i} x_j]$. Hence, this game falls naturally into the framework of Definition 1.22 and (1.31). ◁

By similar arguments as in Proposition 1.19, we can prove the following:

**Proposition 1.24** *Under the same assumptions as in Proposition 1.19, $x^* \in C(x^*)$ is a Nash equilibrium if and only if it satisfies the VI*

$$\langle F(x^*), x - x^* \rangle_{\mathbb{R}^d} \geq 0, \quad \forall\, x \in C(x^*), \tag{1.33}$$

*where*

$$F(x) = (-D_{x_1} U_1, \cdots, -D_{x_m} U_m) : \mathbb{R}^d \to \mathbb{R}^d, \quad d = \sum_{i=1}^m d_i, \quad C(x) = \prod_{j=1}^m K_j(x).$$

There is an important difference between the VI (1.29) and the VI (1.33). In (1.29) the closed convex set $C$ is given and is independent of the solution of the VI. On the other hand, in (1.33) the closed convex set $C(x^*)$ is determined by the solution itself, so it is part of the unknowns as well. Variational inequalities such as (1.33) are called quasi-variational inequalities and play an important role in mathematical modelling in various applications, game theory among others. Their theory presents more difficulties than the theory of standard VIs of the form (1.29).

### 1.2.3 Optimal Stopping: The American Options or Real Options

An important class of applications of VIs is in the theory of optimal stopping. This finds important applications in management science and finance, among others in the study of American options in finance or the theory of real options which is a common tool in the valuation of risky projects or irreversible investment in management science.

As a first approach to this field, we present here a brief introduction to the theory of American or real options and its connection to VIs. Consider an asset or cash flow, whose (random) value $Y_s$ at time $s \geq \tau$ follows the geometric Brownian motion (GBM) process

$$\begin{aligned} dY_s &= \mu Y_s ds + \sigma Y_s dW_s, \quad s \geq \tau, \\ Y_\tau &= y, \end{aligned} \tag{1.34}$$

where $(W_s)_{s \geq 0}$ is a standard Wiener process, and $\mu, \sigma$ are parameters of the model.

We now introduce two related problems both of importance in finance and management science:

$P_1$. We wish to price, at time $\tau$, an American type option on an underlying asset whose value at time $t$ is given by $Y_t$. This contract can be exercised at any stopping time $\mathfrak{s} \in [\tau, T]$. If exercised at $\mathfrak{s}$, the pay-off from this option will be equal to $\psi(Y_{\mathfrak{s}})$, where $\psi$ is a given function characterizing the option; in the case of the American put $\psi(Y_{\mathfrak{s}}) = (K - Y_{\mathfrak{s}})^+$. An important question in financial mathematics is to determine the value of this option at time $\tau$, given the value of the underlying at this time $Y_\tau = y$ and find a rule for the optimal exercise time (see, e.g., Karatzas et al. (1998), Oksendal (2013)).

$P_2$. We assume that (1.34) is a good model for the cash flow an entrepreneur will enjoy while participating in a risky business venture until a chosen stopping time $\mathfrak{s} \in [\tau, T]$. The business venture will offer the entrepreneur an intertemporal pay-off during the period $[\tau, T]$, which of course also involves the cost of running the business, and a final endowment $\psi(Y_{\mathfrak{s}})$ at the termination time of the project $\mathfrak{s}$, both depending on the process $Y := (Y_s)_{s \geq \tau}$. An important question is to find the value of the project and an appropriate time $\mathfrak{s}$ to exit the project. This situation resembles an American option and is called a real option in the management literature (see, e.g., Trigeorgis and Tsekrekos (2018)).

In the above, by stopping time $\mathfrak{s}$ we mean a random time with the property $\{\omega : \mathfrak{s}(\omega) \leq t\} \in \mathscr{F}_t$, where $\mathscr{F}_t$ is the $\sigma$-algebra generated by the Wiener process up to time $t$. This essentially models the fact that the decision to exercise at time $\mathfrak{s}$, before $t$, can only be based on the information about the process that is available up to time $t$, i.e., the past of the process. In both cases $P_1$ and $P_2$, it makes perfect sense to consider either the finite horizon case, $T < \infty$, or the infinite horizon (perpetual) case, $T = \infty$.

Problems $P_1$ and $P_2$, though different in nature, are very similar in terms of modelling and methodology by which they are solved. Needless to say, they both reduce to VIs!

Clearly, the value of the option, either for $P_1$ or for $P_2$, is equal to

$$V(\tau, y) = \sup_{\mathfrak{s} \in \mathscr{T}_{\tau,T}} \mathbb{E}\left[ \int_{\tau}^{\mathfrak{s}} e^{-r(s-\tau)} f(Y_s) ds + e^{-r(\mathfrak{s}-\tau)} \psi(Y_{\mathfrak{s}}) \middle| Y_\tau = y \right], \quad (1.35)$$

where $\mathscr{T}_{\tau,T}$ is the set of stopping times $\mathfrak{s} \in [\tau, T]$, $r > 0$ is an appropriate discount rate, $f$ models the intertemporal pay-offs (including running cost), and $\psi$ is the function modelling the final pay-off. A common variant of this problem is to use as stopping time $\min(\mathfrak{s}, T_D)$, where $T_D$ is the first exit time from an acceptable interval of values for $Y$, which can model options with barrier features, i.e., options that are invalidated when the underlying $Y$ hits the boundary of the region.

**Finite Horizon Problems**

Using standard tools from stochastic analysis (see, e.g., Oksendal (2013) or Karatzas et al. (1998)), it is possible to show that the value function $V$ satisfies the following conditions:[9]

$$\begin{aligned} &\frac{\partial V}{\partial \tau} + \mu y \frac{\partial V}{\partial y} + \frac{1}{2}\sigma^2 y^2 \frac{\partial^2 V}{\partial y^2} - rV + f(y) < 0 \quad \text{if} \quad V(\tau, y) - \psi(y) = 0, \\ &\frac{\partial V}{\partial \tau} + \mu y \frac{\partial V}{\partial y} + \frac{1}{2}\sigma^2 y^2 \frac{\partial^2 V}{\partial y^2} - rV + f(y) = 0 \quad \text{if} \quad V(\tau, y) - \psi(y) > 0, \end{aligned} \quad (1.36)$$

with the above, for the time being, assumed to hold pointwise, i.e., for every $(\tau, y)$. The economic interpretation of (1.36) is quite simple: The quantity on the LHS of (1.36) corresponds to the instantaneous rate of change of the discounted pay-off process $(e^{-rs} V(s, Y_s))_{s \geq \tau}$ while not exercised, with the term $f$ corresponding to the rate of accrual of the intertemporal pay-off. When this quantity gets strictly negative, the holder starts losing value by keeping the option unexercised, hence exercises, and gets the instantaneous value $V = \psi$. Otherwise, it is better to keep the option unexercised, as the value of the option $V > \psi$, the instantaneous pay-off at exercise.

The problem (1.36) has to be complemented with:

- Final/initial[10] conditions related to the behaviour of the value of the option at expiry $\tau = T$, i.e., $V(T, y) = \psi(y)$ and
- Boundary conditions related to the behaviour of the value of the option in case the underlying value starts at some extreme values, for example, $y \to 0$ and

---

[9] In (1.36) we may set $\mu = r$ if we work under the risk-neutral measure.

[10] As we shall see a simple transformation of the form $t = T - \tau$ can turn the final condition to an initial condition where one determines the value of the option at $t = 0$.

$y \to \infty$ or ($y \to \underline{y}$ and $y \to \bar{y}$ for suitable finite values $\underline{y}$, $\bar{y}$ related to the dynamics of the process $Y$, or the boundary of its inactivation region in the case of barrier options), e.g., $V(\tau, \underline{y}) = K_1$ and $V(\tau, \bar{y}) = K_2$ for suitable $K_1, K_2$.

It is common to use the transformation to a new time variable $t = T - \tau$, counting the time left to expiry at the time $\tau$, when the option is to be valued, and consider $V$ as a function of $t$. In terms of this new variable $V_\tau = -V_t$, multiplying (1.36) by a minus sign (for reasons to become apparent in Chap. 2), and defining the operator $A$ by

$$V \mapsto AV := -\frac{1}{2}y^2 V_{yy} - \mu y V_y + rV, \tag{1.37}$$

we see that (1.36) assumes the equivalent form

$$\begin{cases} V_t + AV - f \geq 0, \quad V - \psi \geq 0, \\ \quad (V_t + AV - f)(V - \psi) = 0 \end{cases} \iff \begin{cases} V_t + AV - f > 0, \text{ in } \mathscr{C} = \{V = \psi\}, \\ V_t + AV - f = 0, \text{ in } \mathscr{C}^c = \{V > \psi\} \end{cases}$$
$$\iff \min\{V_t + AV - f, V - \psi\} = 0, \tag{1.38}$$

with the above assumed to hold pointwise in $(t, y)$ and subject to the initial and boundary conditions

$$\begin{aligned} V(0, y) &= \psi(y), \quad y \in [\underline{y}, \bar{y}], \\ V(t, \underline{y}) = K_1, \quad V(t, \bar{y}) &= K_2, \quad t \in [0, T], \end{aligned} \tag{1.39}$$

for suitable choice of $\underline{y}, \bar{y}, K_1, K_2$.

The set $\mathscr{C} = \{V = \psi\} = \{(t, y) : V(t, y) = \psi(y)\}$ is called the contact set and consists of all these points where the solution $V$, which is always above the "obstacle" $\psi$ (see the first formulation in (1.38)), touches the obstacle. The stopping time at which the process $(Y_t)_{t \in [0,T]}$ first enters $\mathscr{C}$ is when the holder stops and exercises the option, so it is often called the exercise region in this literature. This corresponds to a boundary condition which, contrary to the one in (1.39), is not set a priori but depends on and is determined by the unknown solution $V$; this is called a free boundary condition.

If we perform the change of variables to a new function $U$ such that

$$U(t, x) = e^{-\beta x} V(t, x), \quad \text{for } x = \ln(y), \quad \beta = -\frac{\mu}{\sigma^2}, \tag{1.40}$$

it is easy to verify that $U$ satisfies (1.38) for the operator $A_H$, defined by

$$U \mapsto A_H U := -U_{xx} + k^2 U, \quad \text{where } k^2 = r + \frac{\mu}{\sigma^2} > 0, \tag{1.41}$$

and $f$, $\psi$ replaced by $\bar{\psi}(x) = e^{-\beta x}\psi(x)$, and similarly for $\bar{f}$, subject to the initial and boundary conditions (1.39) (expressed in terms of $U$ and $x$).

**Infinite Horizon Projects: Perpetuities**

Another version of the same problem is to consider that the expiry date of the option is too long, i.e., $T \to \infty$. Contracts of this type are called perpetuities and can be quite common in certain markets or can even be used to study the valuation of more general business decisions as, e.g., in real options theory. In this case, the value of the option is provided as in (1.35), but now the supremum is not taken over the set $\mathcal{T}_{\tau,T}$, but rather $\mathcal{T}_\tau$ the set of stopping times $\mathfrak{s} \in [\tau, \infty)$. In this case the value of the option does not depend on $\tau$. A simple way to see this is by using the Markov property for the process $Y := (Y_s)_{s\geq\tau}$. Since $\mathfrak{s} \geq \tau$, we express $\mathfrak{s} = \bar{\mathfrak{s}} + \tau$, where now $\bar{\mathfrak{s}} \geq 0$ is a stopping time. In terms of the new formulation, the value of the option can be expressed as

$$\bar{V}(y) = \sup_{\bar{\mathfrak{s}}\geq 0} \mathbb{E}\left[ \int_0^{\bar{\mathfrak{s}}} e^{-rs} f(\bar{Y}_s)ds + e^{-r\bar{\mathfrak{s}}}\psi(\bar{Y}_{\bar{\mathfrak{s}}}) \middle| \bar{Y}_0 = y \right], \tag{1.42}$$

where $\bar{Y} := (\bar{Y}_{\bar{t}})_{\bar{t}\geq 0}$ is a new process satisfying the same dynamics as (1.34) but starting at $\bar{t} = 0$. This is essentially the same process as (1.34) but shifted in time, i.e., it is essentially $Y_{s-\tau}$. Because of the Markov property for the process $Y$, the value function for the perpetual American option does not depend on $\tau$ (equiv. $t$).

Dropping the bar for ease of notation, we can see that the value function $y \mapsto V(y)$ for the perpetual American put satisfies the linear complementarity problem

$$\begin{cases} AV - f \geq 0, \quad V - \psi \geq 0, \\ \quad (AV - f)(V - \psi) = 0 \end{cases} \iff \begin{cases} AV - f > 0, \text{ in } \mathcal{C} = \{V = \psi\}, \\ AV - f = 0, \text{ in } \mathcal{C}^c = \{V > \psi\} \end{cases}$$
$$\iff \min\{AV - f, V - \psi\} = 0, \tag{1.43}$$

assumed to hold pointwise in $y$, subject only to the boundary conditions in (1.39), that is,

$$V(\underline{y}) = K_1, \quad V(\bar{y}) = K_2. \tag{1.44}$$

Using the change of variables (1.40), the problem reduces to (1.43) for the constant coefficient operator $A_H$.

**Basket Options**

The above modelling exercise can be extended for the case of basket options, i.e., for options whose pay-off depends on $N > 1$ underlying assets, whose values are collected in a vector-valued process $Y = (Y_t)_{t\geq 0}$ with $Y_t = (Y_{1,t}, \cdots, Y_{N,t})$. While the analysis can be very general, let us consider here the specific case where the log prices $X_t = (X_{1,t}, \cdots, X_{N,t})$, $X_{i,t} = \ln Y_{i,t}$, satisfy the multidimensional Itō process

$$\begin{aligned} dX_{i,s} &= \mu_i(X_s)ds + \sum_{j=1}^{m} \sigma_{ij}(X_s)dW_j(s), \quad s \geq \tau, \\ X_{i,\tau} &= x_i, \qquad i = 1, \cdots, N, \end{aligned} \tag{1.45}$$

where now the stochasticity in the market is introduced by $m$ stochastic factors $W(s) = (W_1(s), \cdots, W_m(s))$ modelled as a standard $m$-dimensional Wiener process[11] and an American option in this market whose pay-off is as in (1.35), where now $f, \psi$ depend on the vector $X$. An example for the pay-off function $\psi$ could be $\psi(x_1, \cdots, x_N) = \max(K - \sum_{i=1}^{N} w_i e^{x_i}, 0)$, for prescribed weights $w_i \geq 0, \sum_{i=1}^{N} w_i = 1$, corresponding to a basket put option. The value function $V$ is now a function of $t = T - \tau$ and $x = (x_1, \cdots, x_N)$, and using a multidimensional analogue of Itō's lemma (see, e.g., Oksendal (2013)) can be shown to satisfy a linear complementarity problem of the form (1.38) for the operator

$$U \mapsto AU := -LU + rU := -\sum_{i,j=1}^{N} a_{ij}(x) \frac{\partial^2 U}{\partial x_i \partial x_j} - \sum_{i=1}^{N} \mu_i(x) \frac{\partial U}{\partial x_i} + rU, \tag{1.46}$$

where $\mathscr{A} := (a_{ij})_{i=1,\cdots,N}^{j=1,\cdots,N} = \frac{1}{2}\sigma\sigma^T \in \mathbb{R}^{N\times N}$ with $\sigma = (\sigma_{ij})_{i=1,\cdots,N}^{j=1,\cdots,m} \in \mathbb{R}^{N\times m}$. The operator $L$ defined in (1.46) can be recognized as the generator of the Itō process (1.45) (see, e.g., Oksendal (2013)). If $\mu_i(x) = \mu_i x_i$ (see, e.g., Agliardi and Agliardi (2023)), upon suitable transformations of the independent and the dependent variables, i.e., upon setting $U(t, x) = \exp(\langle \nu, x\rangle)V(t, x)$ for a suitable vector $\nu \in \mathbb{R}^N$, it can be shown that $U$ satisfies a linear complementarity problem of the form (1.38) for the operator

$$U \mapsto \bar{A}U(x) := -\sum_{i,j=1}^{N} a_{ij} \frac{\partial^2 U}{\partial x_i \partial x_j} + k^2 U, \quad \text{where } k^2 = r + \frac{1}{2}\nu^T \mathscr{A} \nu \geq 0, \tag{1.47}$$

where upon defining the new variables $z$ in terms of $x = \mathfrak{D} z$, where $\mathscr{A} = \mathfrak{D}\mathfrak{D}^T$ (Cholesky decomposition), and the function $\bar{U}(z) = U(\mathfrak{D}z)$ satisfies a linear

[11] Note that the correlations between the assets are introduced here in terms of the covariance factors $\sigma_{ij}$ rather than in terms of correlations between the various components of the Wiener process.

complementarity problem of the form (1.38) in terms of the operator

$$\bar{U} \mapsto \hat{A}\bar{U}(z) := -\frac{1}{2}\sum_{i=1}^{N}\frac{\partial^2 \bar{U}}{\partial z_i^2} + k^2\bar{U} = -\frac{1}{2}\Delta_z \bar{U} + k^2\bar{U}, \tag{1.48}$$

with the operator $\Delta_z$ called the Laplacian operator.

Perpetual type basket options are also popular financial instruments. Using an analysis similar to that for their single underlying counterparts, we can see that the value of a perpetual basket option depends only on the initial state $y$ (or $x$) so that the value function satisfies (1.43) with the operator $A$ replaced by any of the multidimensional operators introduced here.

Initial and boundary conditions are also required. The generalization of the initial condition is straightforward. Concerning the boundary condition, we now assume that all possible values for $y$ (or $x$) form a suitable subset $D \subset \mathbb{R}^N$, with boundary $\partial D$. Then setting up boundary conditions means specifying the behaviour of $V$ (or $U$) on $\partial D$. For example, in the case of perpetual basket option, an example of a boundary condition (called inhomogeneous Dirichlet boundary condition) can be

$$V(x) = f(x), \quad \forall\, x \in \partial D, \quad \text{for } f : \partial D \to \mathbb{R}, \text{ given.} \tag{1.49}$$

This is a natural extension in higher dimensions of (1.44).[12]

**American Options and Variational Inequalities**
To recap, the valuation of American options, possibly under suitable transformations, reduces to linear complementarity problems of the general form:

Finite horizon:

$$u_t + Au - f \geq 0, \quad u \geq \psi, \quad (u_t + Au - f)(u - \psi) = 0. \tag{1.50}$$

Infinite horizon:

$$Au - f \geq 0, \quad u \geq \psi, \quad (Au - f)(u - \psi) = 0. \tag{1.51}$$

These are subject to suitable boundary conditions in both cases and initial condition in the finite horizon case, where $A$ is a suitable operator (e.g., as in (1.37), (1.41), (1.46), etc.).

[12] If $D = [\underline{y}, \bar{y}]$, then $\partial D = \{\underline{y}, \bar{y}\}$.

Linear complementarity problems of the form (1.50) or (1.51) can be expressed in terms of (infinite dimensional)[13] VIs.

**Proposition 1.25** *Upon defining*

$$K = \{v \ : \ v(x) \geq \psi(x) \ \forall x \in D, \ \text{where } v \text{ is subject to } (1.49)\,\}, \tag{1.52}$$

*which is a convex set, and assuming that* $u, \psi$ *are continuous and* $Au$ *integrable, an equivalent formulation for the infinite horizon problem* (1.51) *is in terms of the VI*

$$\text{Find } u \in K : \ \langle Au - f, v - u \rangle := \int_D (Au - f)(v - u)dx \geq 0, \ \ \forall \ v \in K, \tag{1.53}$$

*subject to suitable initial and/or boundary conditions.*

A similar result follows for the finite horizon problem, upon replacing $Au - f$ by $u_t + Au - f$ and $K$ by a suitable replacement (see Sect. 2.7.2).

***Proof*** Let $u$ satisfy (1.51). We easily see that for any $v \in K$,

$$\begin{aligned}\int_D (Au - f)(v - u)dx &= \int_{\{u>\psi\}} \underbrace{(Au - f)(v - u)dx}_{=0 \ (1.51)} + \int_{\{u=\psi\}} \underbrace{(Au - f)}_{>0 \ (1.51)} \underbrace{(v - u)}_{=v-\psi} dx \\ &= \int_{\{u=\psi\}} \underbrace{(Au - f)}_{>0 \ (1.51)} \underbrace{(v - \psi)}_{\geq 0 \ v\in K} dx \geq 0.\end{aligned} \tag{1.54}$$

Conversely, let $u$ satisfy (1.53). For any positive function $\phi$ satisfying homogeneous Dirichlet boundary conditions, $v = u + \phi \in K$. Using this $v$ in the VI, we obtain

$$\int_D (Au - f)\phi dx \geq 0, \ \ \forall \ \phi > 0 \implies Au - f \geq 0. \tag{1.55}$$

Assuming that $u$ and $\psi$ are continuous, the set $\mathscr{C}^c = \{x \in I \ : \ u(x) > \psi(x)\}$ is an open set meaning that at any $x \in \mathscr{C}^c$ and any function $\phi > 0$ that may be localized

[13] Since $u \mapsto Au$ is a mapping between function spaces (infinite dimensional spaces), (1.53) is an example of infinite dimensional VIs.

around $x$, it holds that $v_{\pm}(x) := u(x) \pm \epsilon\phi(x) > \psi(x)$, i.e., $v_{\pm} \in K$ for some $\epsilon > 0$ sufficiently small. This arbitrary function $\phi > 0$ must be localized in the set $\mathscr{C}^c$. The fact that $\phi$ vanishes outside the set $\mathscr{C}^c$ will be denoted by $\phi \in C_c(\mathscr{C}^c)$. We now use this function $v_{\pm}$ in the VI to get (using the fact that $\phi$ is localized in $\mathscr{C}^c$),

$$\pm \int_I (Au - f)\phi dx = \pm \int_{\mathscr{C}^c} (Au - f)\phi dx \geq 0$$

$$\Longrightarrow \int_{\mathscr{C}^c} (Au - f)\phi dx = 0, \;\; \forall\, \phi \in C_c(\mathscr{C}^c) \Longrightarrow Au - f = 0 \text{ in } \mathscr{C}^c. \tag{1.56}$$

Combining (1.55) and (1.56), we obtain (1.51). □

We close this section with one of the very few examples in which the VI can be solved explicitly.

**Example 1.26 (The American Perpetual Put)** Consider the case of the American perpetual put in the case where the underlying follows the GBM model (1.34) for $\mu = r$, with pay-off structure $\psi(y) = (K - y)^+$, where $K$ is the option's strike. A "solution" to the linear complementarity problem (1.43) can be obtained in closed form as

$$V(y) = \begin{cases} K - y, & y \leq y_*, \\ \frac{K}{1+\gamma}\left(\frac{y}{y_*}\right)^{-\gamma}, & y > y_*, \end{cases} \quad \text{where } y_* = \frac{K\gamma}{1+\gamma}, \;\; \gamma = \frac{2r}{\sigma^2}. \tag{1.57}$$

The interpretation of this solution is that as long as $Y_t > y_*$ we do not exercise the option, and we exercise the first time that $Y_t$ falls below the threshold $y_*$. In this problem, the contact region $\mathscr{C} = \{y \leq y_*\}$. The reader may verify, as an easy calculus exercise, that $V(y) - \psi(y) > 0$ in $\mathscr{C}^c$, whereas $V(y) - \psi(y) = 0$ in $\mathscr{C}$. Moreover, this is a nice example to experiment with the choice of boundary conditions and the various variable transformations leading to (1.43) for alternative operators.

Observe that $V$ is $C^1$ everywhere but fails to be $C^2$ at $y = y_*$. This affects formulation (1.51) but not (1.53). ◁

## References

Agliardi E, Agliardi R (2023) Pricing multidimensional American options. Int J Financ Stud 11(1):51

Auchmuty G (1989) Duality algorithms for nonconvex variational principles. Numer Funct Anal Optim 10(3–4):211–264

Auslender A, Teboulle M (2000) Lagrangian duality and related multiplier methods for variational inequality problems. SIAM J Optim 10(4):1097–1115

Correa JR, Stier-Moses NE (2011) Wardrop equilibria. Encyclopedia of Operations Research and Management Science Wiley
Cournot AA (1838) Recherches sur les principes mathématiques de la théorie des richesses. L. Hachette
De Palma A, Nesterov Y, et al (1998) Optimization formulations and static equilibrium in congested transportation networks. Preprint
Ekeland I (2006) The best of all possible worlds: mathematics and destiny. University of Chicago Press
Facchinei F, Pang JS (2003) Finite-dimensional variational inequalities and complementarity problems. Springer
Fukushima M (1996) Merit functions for variational inequality and complementarity problems. Nonlinear Optim Appl, 155–170
Goh C (2002) Duality in optimization and variational inequalities, vol 2. CRC Press
Karatzas I, Shreve SE, Karatzas I, Shreve SE (1998) Methods of mathematical finance, vol 39. Springer
Kravvaritis DC, Yannacopoulos AN (2020) Variational Methods in Nonlinear Analysis. De Gruyter
Larsson T, Patriksson M (1994) A class of gap functions for variational inequalities. Math Programm 64:53–79
Mazalov V (2014) Mathematical game theory and applications. Wiley
Nagurney A (2013) Network economics: A variational inequality approach, vol 10. Springer Science & Business Media
Nash J (1951) Non-cooperative games. Ann Math, 286–295
Oksendal B (2013) Stochastic differential equations: an introduction with applications. Springer Science & Business Media
Petracou EV, Yannacopoulos A (2013) Decision theory under risk and applications in social sciences: Ii. game theory. Math Model Multidiscip Appl, 421–447
Trigeorgis L, Tsekrekos AE (2018) Real options in operations research: A review. Eur J Oper Res 270(1):1–24
Wardrop JG (1952) Some theoretical aspects of road traffic research. Proc Inst Civil Eng 1(3):325–362

# Chapter 2
# Analysis

*There is nothing more practical than a good theory.*

*Kurt Lewin*

*You are entering a world of pain ...*

*John Goodman as Walter Sobchak, in The Big Lebowski (dir. Joel and Ethan Cohen)*

## 2.1 Introduction

In this chapter we introduce the theory of variational inequalities (VIs) focusing on results related to their well-posedness, addressing questions such as existence, uniqueness, and other properties of solutions. Even though this may appear too theoretical, it is an important step in any modelling endeavour, as a model is usually a mental artifice reflecting the viewpoint of the researcher concerning a given phenomenon, using the language of mathematics. However, there is no guarantee of consistency of this mental exercise, i.e., that the final mathematical object reflecting all the modeller's view (in our case the VI) is a well-defined object. This is exactly where the theoretical approach enters, providing in terms of well-posedness results the required consistency check for the modelling procedure and at the same time revealing important qualitative aspects of the solution, even if the solution cannot be expressed in closed form (as is often the case). Moreover, as we shall see, some of these theoretical approaches can help in the development of numerical methods for the resolution of VIs.

The chapter consists of two main parts. The first one treats finite dimensional problems (i.e., problems of the form (1.1) with $F : \mathbb{R}^d \to \mathbb{R}^d$), while the second one treats infinite dimensional problems (i.e., problems of the form (1.1) but now $F : X \to Y$, where $X, Y$ are infinite dimensional vector spaces—typically function



A. E. Tsekrekos, A. N. Yannacopoulos, *Variational Inequalities in Management Science and Finance*, EURO Advanced Tutorials on Operational Research,
https://doi.org/10.1007/978-3-031-74119-7_2

spaces, and $\langle \cdot, \cdot \rangle$ is understood as a duality pairing[1] between $X$ and $Y$). To connect with the hands-on examples treated in Chap. 1, the problems of transportation networks and Nash equilibria introduced in Sects. 1.2.1 and 1.2.2 correspond to finite dimensional problems,[2] whereas the problems of American options (both on a single or on multiple assets) introduced in Sect. 1.2.3 correspond to infinite dimensional problems. The need for separate treatment lies on the fact that the methods required for the study of the finite and the infinite dimensional case are quite different, with the latter often needing technicalities, pertaining to the infinite dimensional feature of the spaces $X, Y$. On the other hand, there is the need to study both cases, as the importance of the examples introduced in Chap. 1 indicates. Detailed examples will be studied, motivated by the model problems introduced in Chap. 1.

## 2.2 Important Preliminaries

### 2.2.1 *Hilbert Spaces*

For those readers with a less technical education, let us recall the concept of a Hilbert space:

**Definition 2.1** $X$ is a Hilbert space, if it is a complete normed linear space with norm $\|\cdot\|$ generated by an inner product $\langle \cdot, \cdot \rangle$ as $\|x\|^2 = \langle x, x \rangle$.

Recall that a space is complete under a norm if any Cauchy sequence in $(X, \|\cdot\|)$ converges in $(X, \|\cdot\|)$.[3] Moreover, note that the concept of the inner product is very general: Any positive, symmetric, and bilinear mapping $\langle \cdot, \cdot \rangle : X \times X \to \mathbb{R}$ will qualify as an inner product as long as $\langle x, x \rangle = 0$ if and only if $x = 0$.

[1] A duality pairing between two spaces $X$ and $Y$ is a non-degenerate bilinear map $\langle \cdot, \cdot \rangle : Y \times X \to \mathbb{R}$.

[2] There may of course be versions of these problems requiring infinite dimensional VIs, e.g., in the case of a transportation network in continuous space or of more complicated strategy spaces or mixed strategies, respectively.

[3] A Cauchy sequence $(x_n)_{n\in\mathbb{N}}$ is a sequence such that $\|x_n - x_m\| \to 0$ as $n, m \to \infty$. A sequence $(x_n)_{n\in\mathbb{N}} \subset X$ converges in $X$, if there exists $x \in X$ such that $\|x_n - x\| \to 0$ as $n \to \infty$.

**Do Remember 2.2** *For any Hilbert space, X, the following hold:*

$$
\begin{aligned}
&\|x\|^2 := \langle x, x\rangle, \quad \forall\, x \in X, \\
&|\langle x_1, x_2\rangle| \le \|x_1\|\,\|x_2\|, \quad \forall\, x_1, x_2 \in X, \quad \textit{(Cauchy-Schwarz)}, \\
&\|x_1 + x_2\|^2 + \|x_1 - x_2\|^2 = 2\|x_1\|^2 + 2\|x_2\|^2, \quad \forall\, x_1, x_2 \in X.
\end{aligned}
\tag{2.1}
$$

The Cauchy-Schwarz inequality becomes an equality if and only if there exists $\lambda \in \mathbb{R}$ such that $x_1 = \lambda x_2$. The last identity is called the parallelogram law and can be used to check whether a norm on $X$ is compatible with a Hilbert space structure on this space.

**Example 2.3** Hilbert spaces may be finite or infinite dimensional.

1. A standard example of a finite dimensional Hilbert space can be $\mathbb{R}^d$, equipped with the standard Euclidean inner product $\langle x, z\rangle = x \cdot y := \sum_{i=1}^{d} x_i z_i$ for any elements $x = (x_1, \cdots, x_d)$ and $z = (z_1, \cdots, z_d)$ in $\mathbb{R}^d$, which is connected with the standard Euclidean norm $\|x\| = (\sum_{i=1}^{d} |x_i|^2)^{1/2}$. Of course, the same set may be equipped with different inner products such as, for example, $\langle x, z\rangle_w = \sum_{i=1}^{d} w_i x_i z_i$ for a suitable weight vector $w = (w_1, \cdots, w_d) > 0$ (meaning that $w_i > 0$ for all $i = 1, \cdots, d$), which is connected with the weighted Euclidean norm $\|x\|_* = (\sum_{i=1}^{d} w_i |x_i|^2)^{1/2}$. It is important to know that the same space $\mathbb{R}^d$ endowed with a different norm, e.g., $\|x\|_p = (\sum_{i=1}^{d} |x_i|^p)^{1/p}$, $p \in [1, \infty)$, $p \neq 2$, is not a Hilbert space!
2. If $X$ is a finite dimensional Hilbert space, of dimension $d$, then without much loss of generality the reader may assume that $X = \mathbb{R}^d$ for some finite dimension $d$. Indeed, even if $X$ is a more complicated set, in terms of the use of an appropriate basis $\{\phi_i, \,:\, i = 1, \cdots d\} \subset X$, each element $x \in X$ can be expressed as $x = \sum_{i=1}^{d} x_i \phi_i$, where $x_i \in \mathbb{R}$. Hence, since each element $x \in X$ can be identified by an element $(x_1, \cdots, x_d) \in \mathbb{R}^d$, we can always identify $X$ with $\mathbb{R}^d$, equipped either with the Euclidean or any other inner product.[4] We will use the notation $X \simeq \mathbb{R}^d$. As a simple example let $X$ be the space of polynomials of degree $\mathscr{D} \le d - 1$, which can be identified with $\mathbb{R}^d$ using the basis of monomials $\phi_i(s) = s^i$, $i = 0, 1, \cdots, d - 1$.
3. A standard example of infinite dimensional Hilbert space can be the space $L^2(D)$ consisting of square integrable functions $f : D \subset \mathbb{R}^N \to \mathbb{R}$, equipped with the inner product $\langle f, g\rangle := \int_D f(s)g(s)ds$ and the corresponding norm $\|f\|_{L^2(D)} = (\int_D |f(s)|^2 ds)^{1/2}$. The same set can be equipped with different inner products such as, for example, $\langle f, g\rangle_w := \int_D w(s) f(s) g(s) ds$, for a

[4] Possibly dictated by more involved problems.

suitable weight function $w > 0$. The same space, endowed with a different norm, e.g., $\|f\| = \int_D |f(s)|ds$ is not a Hilbert space!

◁

Hilbert spaces, both finite and infinite dimensional, play an important role in the study of VIs. This should be apparent by now after a careful study of the VIs introduced in Sects. 1.2.1 and 1.2.2 for the finite dimensional case (where essentially the Hilbert space $\mathbb{R}^d$ equipped with the Euclidean inner product is used) and Sect. 1.2.3 where essentially the Hilbert space $L^2(D)$ and its corresponding inner product (introduced above) is used.

An important concept (especially for the infinite dimensional case) is that of the (topological) dual space.

**Definition 2.4 (Dual Space)** The dual space of a Hilbert space $X$, denoted by $X'$, is the space of linear bounded (hence also continuous) maps $x' : X \to \mathbb{R}$, where the mapping $x'$ is to be understood in terms of the bilinear duality pairing $\langle \cdot, \cdot \rangle_{X',X} : X' \times X \to \mathbb{R}$ defined by $x'(x) = \langle x', x \rangle_{X',X}$.

$X'$ is a Hilbert space itself when endowed with the, so-called, dual norm

$$\|x'\|_{X'} := \sup_{x \in X} \frac{|x'(x)|}{\|x\|_X} = \sup_{x \in X} \frac{|\langle x', x \rangle_{X',X}|}{\|x\|_X}.$$

**Notation 2.5** *Given a Hilbert space $X$, we will use $\langle \cdot, \cdot \rangle_X$ and $\langle \cdot, \cdot \rangle_{X',X}$ for the inner product on $X$ and for the duality pairing between $X'$ and $X$, respectively, if there is the possibility of confusion. Otherwise we simply use $\langle \cdot, \cdot \rangle$ for both.*

As we will see shortly, the dual space of a Hilbert space may either be identified with the space itself or be understood as a "larger" space.

**Example 2.6** Consider again the case where $X$ is a set of integrable functions. For simplicity, let $D = I$ an interval. In this case, for any functions $f, g$ can set $\langle f, g \rangle = \int_I f(s)g(s)ds$. This integral may make sense even if $f, g$ are not both in $L^2(I)$ but if $g$ is smoother than an $L^2(I)$ function (e.g., $g$ is such that $\frac{dg}{ds} \in C^1(I)$ or $L^2(I)$) and $f = \frac{dF}{ds}$ for some $F \in L^2(I)$ but $f \notin L^2(I)$ necessarily. A simple formal integration by parts argument indicates that $\langle f, g \rangle$ (as defined above) makes perfect sense for such functions. As $f$ and $g$ no longer belong to the same space $X = L^2(I)$, we may not consider $\langle \cdot, \cdot \rangle$, as an inner product in $X = L^2(I)$ but rather as a pairing between a subset of $X_1 \subset X$ (containing $g$) and a function space, larger than $X$ that contains functions $f$ such that $f = \frac{dF}{ds}$ for some $F \in L^2(I)$. The latter can be turned into the dual space $X_1'$. ◁

In infinite dimensional Hilbert spaces it is important to introduce a new kind of convergence associated with the dual space:

**Definition 2.7 (Weak Convergence)** A sequence $(x_n)_{n\in\mathbb{N}} \subset X$ converges weakly in $X$ if there exists some $x \in X$, such that $\langle z, x_n\rangle \to \langle z, x\rangle$ for all $z \in X'$. This is denoted by $x_n \rightharpoonup x$ in $X$.

Naturally, in the above $\langle\cdot,\cdot\rangle$ denotes the duality pairing between $X'$ and $X$ (see Notation 2.5).

Weak convergence is to be contrasted with strong convergence $x_n \to x$, which implies that $\|x_n - x\| \to 0$. Strong implies weak convergence but not the other way round. In finite dimensional spaces the two convergences coincide. As a consequence of the fact[5] that if $x'(x) = \langle x', x\rangle = 0$ for every $x' \in X'$, then $x' = 0$, the weak limit (if it exists) is unique.

**Example 2.8**

1. Let $X = \ell^2 = \{x = (x_k)_{k\in\mathbb{N}} \,:\, \sum_{k=1}^{\infty} |x_k|^2 < \infty\}$, which is a Hilbert space with norm $\|x\|_{\ell^2}^2 = \sum_{k=1}^{\infty} |x_k|^2$ derived by the inner product $\langle x, z\rangle = \sum_{k=1}^{\infty} x_k z_k$. We identify $X'$ with $X$, using hereafter the notation $X' \simeq X$. Consider the sequence $(x_n)_{n\in\mathbb{N}} \subset \ell^2$ defined by $x_n = \delta_{kn}$. Then for any $z \in \ell^2$, we can see that $\langle z, x_n\rangle = \sum_{k=1}^{\infty} z_k\delta_{kn} = z_n \to 0$ as $n \to \infty$, since $\sum_n |z_n|^2 < \infty$. Hence $x_n \rightharpoonup 0$. On the other hand, $\|x_n - 0\|_{\ell^2} = 1$ for all $n \in \mathbb{N}$; hence $x_n \not\to 0$ in $\ell^2$.
2. Let $X = L^2(I)$, $I = (0, 1)$, with $X' \simeq X$, and consider the sequence of functions $(f_n)_{n\in\mathbb{N}} \subset L^2(I)$ defined by $f_n(s) = \sin(2\pi ns)$. Then $f_n \rightharpoonup 0$, but $f_n \not\to 0$ in $L^2(I)$, since $\|f_n - 0\|_{L^2(I)} \to \frac{1}{2}$ as $n \to \infty$.

◁

Weak convergence is a way to reinstate the important result that closed and bounded implies convergence up to subsequences, which is valid in finite dimensional spaces[6] to the case of infinite dimensional spaces. This follows as a particular case of the general Banach-Alaoglu theorem (see, e.g., Brezis (2011)):

**Proposition 2.9** *A bounded sequence in a (infinite dimensional) Hilbert space $X$ has a weakly convergence subsequence in $X$.*

[5] Which can be seen as a consequence of the Hahn-Banach theorem (see, e.g., Brezis (2011)).

[6] This follows by the Heine-Borel theorem.

**Do Remember 2.10** *The norm is a lower semicontinuous function (in terms of weak convergence)*

$$x_n \rightharpoonup x \implies \|x\| \leq \liminf_n \|x_n\|. \tag{2.2}$$

*It is important to note that*

$$x_n \to x \implies \|x_n\| \to \|x\|,$$
$$x_n \rightharpoonup x \not\Longrightarrow \|x_n\| \to \|x\|,$$

*as in the second case* $\lim_n \|x_n\|$ *may not even exist!*

An important result is the celebrated Riesz-Fréchet representation theorem:

**Theorem 2.11 (Riesz-Fréchet)** *Given any* $x' \in X'$, *there exists a unique* $x \in X$ *such that*

$$\langle x', z\rangle_{X',X} = \langle x, z\rangle_X \;\; \forall\, z \in X, \textit{with } \|x\|_X = \|x'\|_{X'}.$$

*Using the above we may define a single-valued map* $\iota : X' \to X$, *called the Riesz isometry by* $\iota(x') = x$.

In finite dimensional cases we usually identify $X$ with $X'$ using this isometry, so for all applications regarding finite dimensional we will identify $X \simeq \mathbb{R}^d \simeq X'$. In infinite dimensions the situation is more delicate, and we may either choose to identify $X$ with its dual using the Riesz isometry, denoting that with $X \simeq X'$, or not. Unless otherwise indicated for most of this book, we will assume that $X' \simeq X$.

The following example is interesting:

**Example 2.12 (Brezis 2011)** Consider the spaces of sequences $x = (x_n)_{n\in\mathbb{N}} \subset \mathbb{R}$,

$$X_0 = \{(x_n)_{n\in\mathbb{N}} : \sum_n |x_n|^2 < \infty\}, \;\; \|x\|^2_{X_0} = \sum_n |x_n|^2, \;\; \langle x, z\rangle_{X_0} = \sum_n x_n z_n,$$

$$X_1 = \{(x_n)_{n\in\mathbb{N}} : \sum_n w_n |x_n|^2 < \infty, \}, \;\; \|x\|^2_{X_1} = \sum_n w_n |x_n|^2, \;\; \langle x, z\rangle_{X_1} = \sum_n w_n x_n z_n,$$

where $(w_n)_{n\in\mathbb{N}} \subset \mathbb{R}$ is a sequence of strictly positive weights. These are both Hilbert spaces with the corresponding norms. Moreover, between any two sequence spaces $X, Y$, with typical elements $x = (x_n)_{n\in\mathbb{N}}$, $y = (y_n)_{n\in\mathbb{N}}$, respectively, consider the pairing

$$\langle\cdot,\cdot\rangle_{Y,X} : Y \times X \to \mathbb{R}, \quad \text{defined by } \langle y, x\rangle_{Y,X} = \sum_n y_n x_n.$$

In terms of this pairing we may identify $X_0' = X_0$, while if $w_n > 1$, then we may identify the dual space of $X_1$ as

$$X_1' = \{(x_n)_{n\in\mathbb{N}} \,:\, \sum_n w_n^{-1}|x_n|^2 < \infty\}, \quad \|x\|_{X_1'}^2 = \sum_n w_n^{-1}|x_n|^2,$$
$$\langle x, z\rangle_{X_1'} = \sum_n w_n^{-1} x_n z_n,$$

a fact that can be seen directly by noting that for any $x' \in X_1'$ and $x \in X_1$, the pairing $\langle x', x\rangle_{X_1', X_1}$ is well defined (simply use the Cauchy-Schwarz inequality). In this case the Riesz mapping is explicit; for any $x' = (x_n')_{n\in\mathbb{N}} \subset X_1'$, we have $\iota(x') = x = (w_n^{-1}x_n')_{n\in\mathbb{N}} \subset X_1$, and we can see show explicitly that this is an isometry, i.e., that $\|\iota(x')\|_{X_1} = \|x'\|_{X_1'}$. Clearly, if $w_n > 1$, it holds that $X_1 \subset X_0 \subset X_1'$. $\triangleleft$

### 2.2.2 *Projection on Closed Convex Sets*

Before embarking on the study of the theory of finite dimensional VIs, we need to introduce an important technical tool, the projection operator on closed convex sets. We emphasize that all the results of this section remain valid also for the infinite dimensional case, so they will also be used in the second part of the chapter as well.

**Definition 2.13 (Projection Operator and the Normal Cone)** Let $X$ be a Hilbert space, with inner product $\langle\cdot,\cdot\rangle$, and $C \subset X$ a closed convex set.

(i) The mapping $x \mapsto \arg\min_{z\in C} \frac{1}{2}\|x-z\|^2$ defines a single-valued mapping $\Pi_C : X \to C$ called the projection operator on $C$.
(ii) The normal cone of $C$ at any $x \in X$ is defined as

$$N_C(x) = \begin{cases} \{w \in X \,:\, \langle w, z-x\rangle \le 0, \;\; \forall z \in C\} \text{ if } & x \in C, \\ \emptyset & \text{otherwise.} \end{cases}$$

The normal cone can be understood as a multi-valued map $N_C : X \to 2^X$, i.e., a map that maps a point $x \in X$ to a set $N_C(x) \subset X$ (see Fig. 2.1a).

Multi-valued maps play an important role in the theory and computation of VIs (see Chap. 3). Another important multi-valued map is the subdifferential.

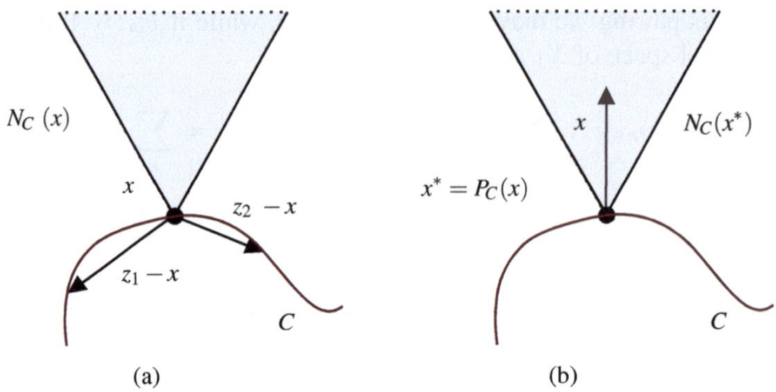


**Fig. 2.1** (**a**) The normal cone $N_c$ at $x \in C$. (**b**) The projection $x^* = P_C(x)$ for $x \notin C$ and its connection with $N_C(x)$. (Courtesy of Dr. G. Domazakis)

**Definition 2.14 (Subdifferential)** Given $\phi : X \to \mathbb{R} \cup \{+\infty\}$, the subdifferential $\partial\phi : X \to 2^X$ is the multi-valued operator (point to set mapping), defined by

$$\partial\phi(x) = \{y \in X \ : \ \phi(x) + \langle y, z - x\rangle \le \phi(z), \ \forall z \in X\}.$$

The subdifferential is the generalization of the derivative for nonsmooth functions, and if $\phi \in C^1$, then $\partial\phi(x) = \{D\phi(x)\}$, i.e., in this case the map is single-valued. The subdifferential finds important applications in nonsmooth optimization since

$$x^* \in \arg\min_{x \in X} \phi(x) \iff 0 \in \partial\phi(x^*).$$

If $X'$ is not identified with $X$, then we can consider $\partial\phi(x) \subset X'$ (not needed here).

**Example 2.15** (**$N_C$ as a Subdifferential**) An important characterization of $N_C$ is as the subdifferential of $I_C$, defined by

$$I_C(x) := \begin{cases} 0 & \text{if } x \in C, \\ +\infty & \text{if } x \notin C. \end{cases}$$

Then, applying the definitions to the convex indicator function $I_C$, we easily see that $N_C = \partial I_C$. ◁

The projection operator and the normal cone satisfy the following properties:

**Proposition 2.16** *Let $C \subset X$ be a closed convex set. Then, the projection operator $\Pi_C : X \to C$:*

*(i) Is single-valued, and $x^* = \Pi_C(x)$ is characterized in terms of the VI*

$$\langle x - x^*, z - x^* \rangle \leq 0, \quad \forall\, z \in C. \tag{2.3}$$

*(ii) Is non-expansive, i.e.,*

$$\|\Pi_C(x_1) - \Pi_C(x_2)\| \leq \|x_1 - x_2\|, \quad \forall\, x_1, x_2 \in X. \tag{2.4}$$

*(iii) $x^* = \Pi_C(x)$ if and only if $x - x^* \in N_C(x^*)$ (see Fig. 2.1(b)).*
*(iv) Satisfies the property*

$$\|\Pi_C(x) - z\|^2 \leq \|x - z\|^2 - \|x - \Pi_C(x)\|^2, \quad \forall\, x \in X,\ z \in C. \tag{2.5}$$

***Proof***

(i) Note that $x^* = \Pi_C(x)$ is the solution of the minimization problem $\min_{z \in C} \|x - z\|^2$. To show the existence of $x^*$, take a minimizing sequence $(x_n)_{n\in\mathbb{N}} \in C$, i.e., a sequence such that $\|x_n - x\|^2 \to \inf_{z\in C} \|z - x\|^2$, and use the parallelogram identity (see (2.1)) for $x - x_n$ and $x - x_m$, to get upon rearrangement

$$\|x_n - x_m\|^2 = 2\Big(\|x - x_n\|^2 + \|x - x_m\|^2 - 2\|x - \frac{1}{2}(x_n + x_m)\|^2\Big), \quad \forall\, n, m \in \mathbb{N}. \tag{2.6}$$

By the convexity of $C$, we have that $\frac{1}{2}(x_n + x_m) \in C$, hence $\|x - \frac{1}{2}(x_n + x_m)\|^2 \geq \inf_{z\in C} \|x - z\|^2$, and passing to the limit as $n, m \to \infty$ we obtain from (2.6) that $\lim_{n,m} \|x_n - x_m\|^2 \to 0$; hence $(x_n)_{n\in\mathbb{N}} \in C$ is Cauchy in $X$ and by the fact that $C$ is closed has a limit $x^* \in C$. By continuity of the norm, $x^*$ is the minimizer. Uniqueness follows once more by the parallelogram law, assuming per contra the existence of two minimizers (or by strict convexity of the square of the norm).

To derive the VI (2.3) consider the minimizer $x^* \in C$ and any $z \in C$. By convexity $x_\epsilon = (1 - \epsilon)x^* + \epsilon z = x^* + \epsilon(z - x^*) \in C$, for any $\epsilon \in (0, 1)$, so that by minimality and the fact that the norm is derived by an inner product,

$$0 \leq \|x^* + \epsilon(z - x^*) - x\|^2 - \|x^* - x\|^2 = \epsilon\langle x^* - x, z - x^*\rangle + \frac{\epsilon^2}{2}\|z - x^*\|^2,$$

which upon dividing by $\epsilon > 0$ and passing to the limit as $\epsilon \to 0^+$ yields (2.3).

$$\langle x^* - x, z - x^* \rangle \geq 0, \quad \forall\, z \in C.$$

(ii) The result follows by the use of the VI, defining the projection. Take any pair $(x_1, x_2) \in X \times X$. Apply (2.3) (multiplied by $-1$), first for $(x, z) = (x_1, \Pi_C(x_2))$ and then for $(x, z) = (x_2, \Pi_C(x_1))$. This yields

$$\langle \Pi_C(x_1) - x_1, \Pi_C(x_2) - \Pi_C(x_1) \rangle \geq 0,$$
$$\langle \Pi_C(x_2) - x_2, \Pi_C(x_1) - \Pi_C(x_2) \rangle \geq 0.$$

Adding these two and rearranging, we get

$$\langle (x_1 - x_2) - (\Pi_C(x_1) - \Pi_C(x_2)), \Pi_C(x_1) - \Pi_C(x_2) \rangle \geq 0,$$

which yields

$$\begin{aligned} \|\Pi_C(x_1) - \Pi_C(x_2)\|^2 &\leq \langle x_1 - x_2, \Pi_C(x_1) - \Pi_C(x_2) \rangle \\ &\leq \|x_1 - x_2\| \|\Pi_C(x_1) - \Pi_C(x_2)\|, \end{aligned}$$

where for the last inequality we used the Cauchy-Schwarz inequality. Dividing both sides by $\|\Pi_C(x_1) - \Pi_C(x_2)\|$ yields the required result.

(iii) If $x^* = \Pi_C(x)$, then it solves the VI (2.3). Upon defining $w = x - x^*$ and recalling the definition of the normal cone yields $w = x - x^* \in N_C$.

(iv) We start by the observation

$$\begin{aligned} \|x - z\|^2 &= \|x - \Pi_C(x) + (\Pi_C(x) - z)\|^2 \\ &= \|x - \Pi_C(x)\|^2 + \|\Pi_C(x) - z\|^2 + 2\underbrace{\langle x - \Pi_C(x), \Pi_C(x) - z \rangle}_{\geq 0 \text{ by } (2.3)} \\ &\geq \|x - \Pi_C(x)\|^2 + \|\Pi_C(x) - z\|^2, \end{aligned}$$

from which we immediately obtain (2.5).

□

**Example 2.17** Let $C = \mathbb{R}^d_+$. Then, $\Pi_C(x) = \max(x, 0) = x^+$, where the positive part of the vector $x$ is understood coordinatewise.

It suffices to show that the above proposal solves the VI (2.3). Take any vector $x = (x_1, \cdots, x_d)$, and consider any $z = (z_1, \cdots, z_d) \in C$, where $z_i \geq 0$. Noting that $s^+ - s = \max(s, 0) - s = s^- := \max(-s, 0) \geq 0$ and that $s^+ s^- = 0$ for any

$s \in \mathbb{R}$, we see that for all $z \in C$,

$$\langle x - x^+, z - x^+\rangle = -\sum_{i=1}^{d} x_i^-(z_i - x_i^+) = -\sum_{i=1}^{d} x_i^- z_i + \sum_{i=1}^{d} x_i^- x_i^+$$
$$= -\sum_{i=1}^{d} x_i^- z_i \leq 0$$

hence solves (2.3). ◁

**Example 2.18** Let $C = \{f \in L^2(D) \ : \ f \geq 0\}$, the subset of square integrable functions $f : D \to \mathbb{R}$ such that $f(s) \geq 0$ a.e. in $D$. This is a closed convex set. For any function $f \in X = L^2(D)$, it holds that $\Pi_C(f) = f^+ := \max(f, 0)$, i.e., the function defined pointwise by $f^+(s) = \max(f(s), 0)$.

Indeed, noting that pointwise $f^+ - f = f^- := \max(-f, 0) \geq 0$ and $f^+ f^- = 0$ and using the properties of the integral,

$$\langle f - f^+, g - f^+\rangle = \int_D (f(s) - f^+(s))(g(s) - f^+(s))ds$$
$$= -\int_D f^-(s)(g(s) - f^+(s))ds$$
$$= -\int_D f^-(s)g(s)ds + \int_D f^-(s)f^+(s)ds$$
$$= -\int_D f^-(s)g(s)ds \leq 0, \ \ \forall \ g \in C$$

hence solves (2.3).

A similar calculation yields that if $C$ is as above but now $f \geq \psi$ in $D$, where $\psi$ is a given function, then $\Pi_C(f) = (f - \psi)^+ + \psi$. ◁

**Remark 2.19**

1. We see from the characterization (2.3) that the projection is directly related to VIs. This is a recurrent theme that will be developed further in the next section and that will be heavily used in the study of the theory and numerical approximation of VIs. The non-expansive property of the projection operator $\Pi_C$ plays an important role in the construction of analytic or numerical schemes for the treatment of VIs.
2. In general, the projection operator $\Pi_C$ is not a linear operator. The linearity of the projection operator holds if $C = E$ is a linear subspace. Moreover, in this case one can show by appropriate choice of test elements $z \in E$ that the corresponding VI reduces to the orthogonality condition $\langle x^* - x, z - x\rangle = 0$ for all $z \in E$.
3. In many important cases the projection operator may not be obtained in closed form as in Example 2.17 above, but it still remains an important conceptual tool, for both analytical and numerical considerations. ⊕

## 2.3 Important Facts Concerning VIs

We will now consider the general problem of a VI of the form

$$x^* \in C \ : \ \langle F(x^*), x - x^* \rangle \geq 0, \ \ x \in C, \tag{2.7}$$

where $C \subset X$ is a closed and convex subset of the Hilbert space $X$ (finite or infinite dimensional) and $F$ is a mapping such that:

- Either $F : X \to X$ , in which case $\langle \cdot, \cdot \rangle$ is interpreted as the inner product in $X$
- Or $F : X \to X'$, in which case $\langle \cdot, \cdot \rangle$ is interpreted as a pairing between the spaces $X'$ and $X$.

As will become clear from Sect. 2.5.1 onwards, the second case can be very important if $X$ is an infinite dimensional space (e.g., a function space or a sequence space). Recall that such problems appear naturally when considering optimal stopping applications (such as American options or real options pricing). We will use the same notation $\langle \cdot, \cdot \rangle$ for both cases and specify the mapping $F$ instead.

### *2.3.1 Connection of VIs with Projection Operators*

We now show the connection of solutions of VIs of the form (2.7), with fixed points of the projection operator $\Pi_C$, as well as with the Wiener-Hopf equation (presented in (2.9) below, see, e.g., Shi (1991), Noor (1993)). We also emphasize that the results of this section hold in any Hilbert space, $X$, finite or infinite dimensional.

**Proposition 2.20** *Define* $F^{\diamond} : X \to X$ *by*

$$F^{\diamond} := \begin{cases} F \text{ if } F : X \to X, \\ \iota F \text{ if } F : X \to X', \end{cases}$$

*where* $\iota$ *is the Riesz isometry defined in Theorem 2.11 .*

*The following are equivalent:*

*(i)* $x^* \in C$ *is a solution of the VI*

$$\langle F(x^*), x - x^* \rangle \geq 0, \ \ \forall x \in C. \tag{2.8}$$

*(ii) For any* $\gamma > 0$, $x^*$ *is a fixed point of the operator* $T_\gamma : C \to C$, *defined by* $T_\gamma = \Pi_C(I - \gamma F^{\diamond})$.

(continued)

**Proposition 2.20** (continued)
*(iii)* $0 \in \gamma F^{\diamond}(x^*) + N_C(x^*)$, *or equivalently* $-\gamma F^{\diamond}(x^*) \in N_C(x^*)$, *for any* $\gamma > 0$.
*(iv) The Wiener-Hopf equation*

$$\gamma F^{\diamond}(\Pi_C(z)) + (I - \Pi_C)(z) = 0 \tag{2.9}$$

*has a solution* $z \in X$ *where*

$$\begin{aligned} x^* &= \Pi_C(z), \\ z &= x^* - \gamma F^{\diamond}(x^*). \end{aligned} \tag{2.10}$$

***Proof*** We consider the two cases separately.

A. Consider the case where $F : X \to X$. We first prove $(i) \iff (ii) \iff (iii)$.
$(i) \implies (ii)$: Let $x^* \in C$ solve (2.8). Pick any $x \in C$ and any $\gamma > 0$. Then,

$$\begin{aligned} &\langle F(x^*), x - x^* \rangle \geq 0 \implies \langle -\gamma F(x^*), x - x^* \rangle \leq 0 \\ \implies &\langle x^*, x - x^* \rangle + \langle -\gamma F(x^*), x - x^* \rangle \leq \langle x^*, x - x^* \rangle, \end{aligned}$$

which can be rearranged as

$$\langle x^* - (x^* - \gamma F(x^*)), x - x^* \rangle \geq 0, \quad \forall\, x \in C. \tag{2.11}$$

Setting $w = x^* - \gamma F(x^*)$, we see that (2.11) is of the form

$$\langle w - x^*, x - x^* \rangle \leq 0, \quad \forall\, x \in C,$$

which upon comparison with the VI defining projections (2.3) implies $x^* = \Pi_C(w)$. Recalling the definition of $w$, we see that

$$x^* = \Pi_C(x^* - \gamma F(x^*)), \tag{2.12}$$

which implies that $x^*$ is a fixed point of the operator $T_\gamma := \Pi_C \circ (I - \gamma F)$. For simplicity, we will denote $T_\gamma = \Pi_C(I - \gamma F)$.
$(ii) \implies (iii)$: Let $x^*$ be a fixed point of the operator $T_\gamma = \Pi_C(I - \gamma F)$. Then,

$$\Pi_C(x^* - \gamma F(x^*)) = x^*,$$

and using Proposition 2.16(ii) this yields

$$(x^* - \gamma F(x^*)) - x^* = -\gamma F(x^*) \in N_C(x^*).$$

$(iii) \implies (i)$: Let $-\gamma F(x^*) \in N_C(x^*)$ for any $\gamma > 0$. Then, by the definition of the normal cone,

$$\langle -\gamma F(x^*), z - x^* \rangle \leq 0, \quad \forall\, z \in C,$$

which implies

$$\langle F(x^*), z - x^* \rangle \geq 0, \quad \forall\, z \in C;$$

hence $x^*$ solves the VI (2.8).

Hence $(i) \iff (ii) \iff (iii)$. It remains to show that $(i) \iff (iv)$.
$(i) \implies (iv)$: Let $x^*$ be a solution of the VI (2.8). Then, by (ii), we have that

$$x^* = \Pi_C(x^* - \gamma' F(x^*)), \quad \gamma' > 0. \tag{2.13}$$

Let $z = x^* - \gamma' F(x^*)$, so that (2.13) reads $x^* = \Pi_C(z)$. We rearrange the definition of $z$ as

$$z = \Pi_C(z) - \gamma' F(\Pi_C(z)) \implies F(\Pi_C(z)) + \frac{1}{\gamma'}(I - \Pi_C)(z) = 0,$$

which is the Wiener-Hopf equation (2.9), for $\gamma = \frac{1}{\gamma'} > 0$.
$(iv) \implies (i)$: Let $z \in X$ be a solution of the Wiener-Hopf equation (2.9). Clearly, upon defining $x^* = \Pi_C(z)$, then (2.9) implies the second equation in (2.10). We will show that $x^* = \Pi_C(z)$ is a solution of the VI (2.8).

The Wiener-Hopf equation (2.9) is rearranged as

$$\gamma F(\Pi_C(z)) = \Pi_C(z) - z. \tag{2.14}$$

We now recall the definition of $\Pi_C(z)$, in terms of the VI (see (2.3)),

$$0 \leq \langle \Pi_C(z) - z, z' - \Pi_C(z) \rangle, \quad \forall\, z' \in C. \tag{2.15}$$

Using (2.14) we see that (2.15) yields

$$0 \leq \langle \gamma F(\Pi_C(z)), z' - \Pi_C(z) \rangle, \quad \forall\, z' \in C, \tag{2.16}$$

which by the fact that $\gamma > 0$ is arbitrary implies that $x^* = \Pi_C(z)$ is a solution of the VI (2.8).

B. Consider the case where $F : X \to X'$. We show the equivalence of (i) and (ii). We first note that (2.8) can be expressed in terms of the Riesz-Fréchet isometry (see Theorem 2.11) as

$$\langle F(x^*), z - x^* \rangle_{X', X} = \langle \iota F(x^*), z - x^* \rangle_X \geq 0, \quad \forall z \in C. \tag{2.17}$$

For $\gamma > 0$, this is equivalent to

$$\langle x - (x - \gamma \iota F(x)), z - x \rangle_X \geq 0, \ \ \forall z \in C \implies x = \Pi_C(x - \gamma(\iota F(x)), \tag{2.18}$$

which is our claim. The other claims follow easily. The proof is complete. □

The connection of the solutions of VIs with fixed points of appropriate operators related to the projection operator on $C$ is very important as it allows us to employ the arsenal of fixed point theory—a well-established and rich mathematical theory—for the study of VIs. This will be extremely useful for both analytic work and the construction of iterative numerical schemes for their solution.

### 2.3.2 Monotonicity and Its Effects

Monotonicity plays an important role in a large part of the theory and numerics of VIs.

**Definition 2.21 (Monotone Mappings)** A mapping $F : X \to X$ is called:

(i) Monotone if

$$\langle F(x_1) - F(x_2), x_1 - x_2 \rangle \geq 0, \ \ \forall \, x_1, x_2 \in X.$$

(ii) Strongly monotone if there exists some $\rho > 0$ such that

$$\langle F(x_1) - F(x_2), x_1 - x_2 \rangle \geq \rho \|x_1 - x_2\|^2, \ \ \forall \, x_1, x_2 \in X.$$

(iii) Pseudo-monotone if

$$\langle F(z), x - z \rangle \geq 0 \implies \langle F(x), x - z \rangle \geq 0, \ \ \forall \, x, z \in X.$$

Clearly a monotone operator is also pseudo-monotone. Moreover, the above definition can be extended for more general types of mappings. For example:

- If $F : X \to 2^X$ is multi-valued (point to set), then replace $F(x)$ by $z \in F(x)$.
- If $F : X \to X'$ or $F : X \to 2^{X'}$, then $\langle \cdot, \cdot \rangle$ is a pairing between $X'$ and $X$.

**Example 2.22** The following are important examples of monotone mappings:

1. Let $F : \mathbb{R}^d \to \mathbb{R}^d$, with $F(x) = Mx$, where $M \in \mathbb{R}^d \times \mathbb{R}^d$ is a positive definite matrix. Then $F$ is monotone.

2. Let $F : X \to X$ be $F(x) = D\phi(x)$, where $\phi$ is a $C^1$ convex function. Then $F$ is monotone. If $\phi$ is strongly convex (i.e., there exists $\rho > 0$ such that $x \mapsto \phi(x) - \frac{\rho}{2}\|x\|^2$ is convex), then $F$ is strongly monotone. This is an important example that holds in both finite and infinite dimensions (in this case $D\phi$ is the Gâteaux derivative of $\phi$).
3. Let $F : X \to 2^X$ be $F = \partial\phi$, the subdifferential of a proper convex function $\phi : X \to \mathbb{R} \cup \{+\infty\}$ (see Definition 2.14). Then $F$ is a (multi-valued) monotone operator.
4. Let $\phi$ be a differentiable pseudoconvex function (see Example 1.6). Then, $F = D\phi$ is a pseudo-monotone operator.

◁

**Definition 2.23 (Coercive Mappings)** A mapping $F : X \to X$ is called:

(i) Coercive if

$$\lim_{\|x\|\to\infty} \frac{\langle F(x), x\rangle}{\|x\|} = +\infty.$$

(ii) Coercive over $C \subset X$ if there exists coercive over $C \subset X$ if there exists $x_0 \in C$ such that

$$\lim_{x\in C,\ \|x\|\to\infty} \frac{\langle F(x), x - x_0\rangle}{\|x\|} = +\infty.$$

The other important notion in the theory of VIs is continuity. There are many possible notions of continuity, which are needed especially in the case when $X$ is of infinite dimension. We start with a fundamental definition:

**Definition 2.24** $F : X \to X$ is called:

(i) Continuous if for any $(x_n)_{n\in\mathbb{N}} \subset X$, such that $x_n \to x$, it holds that $F(x_n) \to F(x)$ (convergence in $X$ in both cases).
(ii) Lipschitz continuous if there exists $c_L > 0$ such that

$$\|F(x_1) - F(x_2)\| \le c_L \|x_1 - x_2\|, \ \ \forall\, x_1, x_2 \in X.$$

If $X$ is finite dimensional, this is pretty much all we need. If $X$ is infinite dimensional, the above will also cover a lot but not all; recall that we may require also a different type of convergence, weak convergence (see Definition 2.7). In this case it makes sense to consider notions of continuity that will work in the case where $x_n \rightharpoonup x$ (weak convergence) in $X$.

The following reformulation of a certain class of VIs is often very useful.

**Proposition 2.25 (Minty's Trick)** *Suppose that $F$ is continuous and pseudo-monotone. Then the VI*

$$x^* \in C \ : \ \langle F(x^*), x - x^* \rangle \geq 0, \ \ \forall\, x \in C \tag{2.19}$$

*is equivalent to the VI*

$$x^* \in C \ : \ \langle F(x), x - x^* \rangle \geq 0, \ \ \forall\, x \in C. \tag{2.20}$$

*The VI* (2.20) *is called Minty's form (MVI)*[7] *of the VI* (2.19).

***Proof*** Suppose that $x^*$ solves (2.19). Then, since $F$ is pseudo-monotone, by the definition of pseudo-monotonicity we obtain (2.20). Note that for this step we do not need the continuity of $F$.

Assume now that $x^*$ solves (2.20), which for notational convenience we rewrite as

$$x^* \text{ solves } \langle F(z), z - x^* \rangle \geq 0, \ \ \forall\, z \in C. \tag{2.21}$$

Consider an arbitrary $x \in C$ and $\epsilon > 0$, and set $z = (1 - \epsilon)x^* + \epsilon z = x^* + \epsilon(x - x^*) \in C$ (by convexity of $C$). Using this choice for $z$ in (2.21), we obtain after rearrangement that

$$\langle F(x^* + \epsilon(x - x^*)), x - x^* \rangle \geq 0, \ \ \epsilon > 0,$$

and passing to the limit as $\epsilon \to 0^+$, using the continuity of $F$ we obtain (2.19). □

Minty's reformulation (sometimes referred to also as Minty's trick) allows us to recover useful properties of the solution set such as, for example, convexity.

**Proposition 2.26** *Let $F$ be continuous and pseudo-monotone. Then the solution set of the VI* (2.8), *$SOL(F, C)$, is closed and convex.*

***Proof*** Let us use the simplified notation $S := SOL(F, C)$, for the solution set of (2.8)

$$S = \{x^* \in C \ : \ \langle F(x^*), x - x^* \rangle \geq 0, \ \ \forall\, x \in C\}.$$

[7] Or sometimes the weak form, but not to be confused with Definition 2.55 or 2.86.

Assume that $S \neq \emptyset$, and consider any two elements $x_1^*, x_2^* \in S$. Using Minty's reformulation $x_i^*$, $i = 1, 2$, satisfy the inequalities

$$\langle F(x), x - x_i^* \rangle \geq 0, \quad \forall\, x \in C, \quad i = 1, 2.$$

We multiply the inequality for $i = 1$ by $\lambda \in (0, 1)$ and the inequality for $i = 2$ by $1 - \lambda$ and add to obtain that $x_\lambda^* = \lambda x_1^* + (1 - \lambda) x_2^* \in C$ satisfies

$$\langle F(x), x - x_\lambda^* \rangle \geq 0, \quad \forall x \in C,$$

i.e., satisfies Minty's form of the VI. By Proposition 2.25 (using the continuity and pseudo-monotonicity of $F$), this implies that

$$\langle F(x_\lambda^*), x - x_\lambda^* \rangle \geq 0, \quad \forall\, x \in C;$$

hence $x_\lambda^* \in S$. This proves the convexity of $S$.

We now show that $S$ is closed. Consider a sequence $(x_n^*)_{n \in \mathbb{N}} \subset S$ such that $x_n^* \to x^*$ in $X$. We need to show that $x^* \in S$. Since $x_n^* \in S$, using Minty's reformulation we have for any $x \in C$ that

$$\langle F(x), x - x_n^* \rangle \geq 0, \quad n \in \mathbb{N}.$$

Passing to the limit as $n \to \infty$, we have that

$$\langle F(x), x - x^* \rangle \geq 0, \quad \forall\, x \in C,$$

so using Minty's trick again we have

$$\langle F(x^*), x - x^* \rangle \geq 0, \quad \forall\, x \in C,$$

and hence $x^* \in S$. This proves that $S$ is closed. □

Using Minty's reformulation we may also provide, under conditions, a saddle point formulation for VIs.

**Proposition 2.27 (Saddle Point Formulation)** *Let $F$ be continuous and pseudo-monotone. Then $x^*$ is a solution of the VI* (2.7) *if and only if the pair $(x^*, x^*)$ is a solution to the saddle point problem for the Lagrangian*

$$(x, y) \mapsto L(x, y) = \langle F(x), y - x \rangle.$$

***Proof*** Let $x^*$ solve (2.7). Then, $L(x^*, y) \geq 0$ for all $y \in C$, while $L(x^*, x^*) = 0$. Using Minty's reformulation $x^*$ solves (2.20); hence $L(x, x^*) \leq 0$. Therefore

$$L(x, x^*) \leq 0 = L(x^*, x^*) \leq L(x^*, x),$$

i.e., $(x^*, x^*)$ is a saddle point for $L$.

Conversely, if $(x^*, x^*)$ is a saddle point for $L$, then it is immediate to see that it satisfies (2.7). □

### *2.3.3 Existence of Solutions: An Abstract Existence Result*

We have so far seen many useful properties of the solution set for VIs; however we have not yet considered the fundamental question:

Do solutions of VIs of the form (2.7) exist and if yes under which conditions?

It turns out that this question can have rather diverse answers depending on whether $X$ is finite or infinite dimensional. In each case one may provide existence results under general conditions, in terms of the celebrated Lions-Stampacchia theorem (see Lions and Stampacchia (1967)).

**Theorem 2.28 (Lions-Stampacchia)** *Let $F : X \to X$ or $F : X \to X'$ be a strongly monotone and Lipschitz continuous operator. Then, the VI* (2.7) *has a unique solution.*

***Proof*** Recall (see Proposition 2.20) that the solvability of (2.8) is reduced to the existence of a fixed point of the operator $T_\gamma := \Pi_C(I - \gamma F^\diamond)$ for some $\gamma > 0$. The strategy of the proof is to show that $T_\gamma$ has a fixed point by using a fixed point theorem. We will use Banach's fixed point theorem (see Theorem 2.107), which is a tool equally valid in finite and infinite dimensional spaces. Then, we will show that for a suitable choice of $\gamma > 0$, the mapping $T_\gamma : C \to C$ is a contraction, so that by Banach's contraction theorem it has a fixed point, which is a solution of (2.8).

Consider first the case $F : X \to X$.

We have that

$$\begin{aligned} \|T_\gamma(x_1) - T_\gamma(x_2)\| &= \|\Pi_C(x_1 - \gamma F(x_1)) - \Pi_C(x_2 - \gamma F(x_2))\| \\ &\leq \|(x_1 - x_2) - \gamma(F(x_1) - F(x_2)\|, \end{aligned} \tag{2.22}$$

where we used the fact that $\Pi_C$ is a non-expansive operator. To further proceed with the RHS, we make the following calculation:

$$\begin{aligned} &\|(x_1 - x_2) - \gamma(F(x_1) - F(x_2))\|^2 \\ &= \langle (x_1 - x_2) - \gamma(F(x_1) - F(x_2)), (x_1 - x_2) - \gamma(F(x_1) - F(x_2))\rangle \\ &= \|x_1 - x_2\|^2 - 2\gamma\langle x_1 - x_2, F(x_1) - F(x_2)\rangle + \gamma^2\|F(x_1) - F(x_2)\|^2 \\ &\leq \|x_1 - x_2\|^2 - 2\gamma\rho\|x_1 - x_2\|^2 + \gamma^2\|F(x_1) - F(x_2)\|^2 \\ &\leq \|x_1 - x_2\|^2 - 2\gamma\rho\|x_1 - x_2\|^2 + \gamma^2 c_L^2\|x_1 - x_2\|^2 \\ &= (1 - 2\gamma\rho + \gamma^2 c_L^2)\|x_1 - x_2\|^2, \end{aligned} \tag{2.23}$$

where in the fourth line we used the strong monotonicity of $F$ (see Definition 2.21(iii) and in the fifth line we used the Lipschitz continuity of $F$ (see Definition 2.24(ii)).

If we choose $\gamma > 0$ so that

$$1 - 2\gamma\rho + \gamma^2 c_L^2 < 1 \implies \gamma < \frac{2\rho}{c_L^2},$$

then combining (2.22) and (2.23) we see that $T_\gamma$ is a contraction map; hence by Banach's contraction mapping theorem admits a fixed point, which is a solution of VI (2.7).

In the case where $F : X \to X'$, we need the following changes in the arguments above:

$$\begin{aligned} \|T_\gamma(x_1) - T_\gamma(x_2)\|^2 &\leq \|(x_1 - x_2) - \gamma(\iota F(x_1) - \iota F(x_2))\|^2 \\ &= \|x_1 - x_2\|^2 - 2\gamma\langle x_1 - x_2, \iota F(x_1) - \iota F(x_2)\rangle + \gamma^2\|\iota F(x_1) - \iota F(x_2)\|^2 \\ &= \|x_1 - x_2\|^2 - 2\gamma\langle F(x_1) - F(x_2), x_1 - x_2\rangle_{X',X} + \gamma^2\|F(x_1) - F(x_2)\|_{X'}^2 \\ &\leq (1 - 2\gamma\rho + \gamma^2 c_L^2)\|x_1 - x_2\|^2, \end{aligned}$$

where we have used the properties of the Riesz isometry $\iota$ (see Proposition 2.11) and the strong monotonicity and Lipschitz continuity of the mapping $F$. From this point onwards the proof proceeds as in the previous case. □

## 2.4 Finite Dimensional Theory

In this section we study the theory of VIs in finite dimensional Hilbert spaces.

**Problem 2.29** Given a finite dimensional Hilbert space $X$, with inner product $\langle \cdot, \cdot \rangle$, a mapping $F : X \to X$, and a closed convex set $C \subset X$, find

$$x^* \in C \; : \; \langle F(x^*), x - x^* \rangle \geq 0, \;\; \forall \, x \in C. \tag{2.24}$$

The VI (2.24) is called finite dimensional because of the fact that $\dim(X) = d < \infty$. Without loss of generality we may consider the case $X = \mathbb{R}^d$ with $\langle \cdot, \cdot \rangle$ the standard Euclidean inner product (see Example 2.3, 2). As we have seen in Sects. 1.2.1 and 1.2.2 of Chap. 1, important classes of problems with important applications in management science reduce to problems in the general form of (2.24).

### *2.4.1 Existence Results*

In this section we will explore the theory of VIs such as (2.24) with a focus on the properties that the function $F$ must satisfy so that problem (2.24) admits solutions. For a comprehensive review, see Harper and Pang (1990).

One may obtain existence results under less restrictive conditions on the mapping $F$ than these required in the abstract existence result, Theorem 2.28, which are imposed so as to guarantee the use of the contraction mapping theorem for $T_\gamma$. Such extensions usually require compactness on the set $C$. In finite dimensional problems (unlike infinite dimensions), compactness is guaranteed as long as $C$ is closed and bounded, hence making the use of fixed point theorems such as the Brouwer fixed point theorem (see Theorem 2.106) easy to implement. In cases where $C$ is not bounded, e.g., in the case where $C = \mathbb{R}_+$, then compactness does not hold, but one may impose extra conditions on $F$ (such as coercivity on $C$) which will do the trick.

**Theorem 2.30** *Let $C$ be a convex and closed set. Assume that one of the following holds:*

*(i) $F$ is continuous and $C$ is compact.*
*(ii) $F$ is continuous and coercive over $C$.*
*(iii) $F$ is continuous and strongly monotone over $C$.*

*Then the VI* (2.8) *admits a solution. Moreover, in case (iii) the solution is unique.*

***Proof*** Recall that the solvability of (2.8) is reduced to the existence of a fixed point of the operator $T_\gamma := \Pi_C(I - \gamma F)$ for some $\gamma > 0$. Note that for all the cases considered $T_\gamma$ is continuous.

(i) It follows directly by application of Brouwer's fixed point theorem.

(ii) By the coercivity of $F$, there exists $x_0 \in C$ such that

$$\lim_{x \in C,\ \|x\| \to \infty} \frac{\langle F(x), x - x_0 \rangle}{\|x\|} = +\infty.$$

This implies that for any $M > 0$ there exists $R > 0$ such that

$$\langle F(x), x - x_0 \rangle \geq M\|x\|, \quad \forall x \ \text{ with } \ \|x\| > R.$$

Hence,

$$\exists\, x_0 \in C \ : \ \langle F(x), x_0 - x \rangle \leq 0, \quad \forall\, x \in C \setminus B(x_0, R),$$

i.e., there is no solution of (2.8) outside the ball $B(x_0, R)$. That means that we can restrict the VI on the set $C \cap \bar{B}(x_0, R)$ which is compact and work as for (i).

(iii) Let $F$ be strongly monotone. Then, by definition there exists $x_0 \in C$ (with $\|x_0\|$ finite) such that

$$\langle F(x) - F(x_0), x - x_0 \rangle \geq \rho\|x - x_0\|^2.$$

Rearranging by bringing $\langle F(x_0), x - x_0 \rangle$ to the other side, dividing by $\|x - x_0\|$, and passing to the limit as $\|x\| \to \infty$, we obtain the coercivity condition. Hence, we have that $F$ is coercive over $C$ and the existence reduces to (ii). Moreover, in this case, consider any two solutions $x_1^*, x_2^*$ of (2.8). Then, by strong monotonicity it holds that

$$\rho\|x_1^* - x_2^*\|^2 \leq \langle F(x_1^*) - F(x_2^*), x_1^* - x_2^* \rangle. \tag{2.25}$$

Since $x_1^*$ solves (2.8) using as test element $x_2^* \in C$ and then reversing the role of $x_1^*, x_2^*$, we have that

$$\langle F(x_1^*), x_2^* - x_1^* \rangle \geq 0,$$
$$\langle F(x_2^*), x_1^* - x_2^* \rangle \geq 0.$$

Combining these with (2.25), we conclude that $\|x_1^* - x_2^*\|^2 \leq 0$ from which we obtain that $x_1^* = x_2^*$, i.e., the solution is unique.

□

**Remark 2.31** Cases (i)–(iii) cannot be directly extended to the infinite dimensional case as their proof relies on the Brouwer fixed point theorem which holds only in finite dimensions. In such cases these results have to be modified and extended using other results such as, for instance, the Schauder fixed point theorem. In such cases the question of compactness has to be carefully considered, as the characterization of compact sets in infinite dimensional spaces is not as straightforward as in the finite dimensional case. In such problems, the notions of weak compactness or of compact embeddings of appropriate subspaces of $X$ into it can be very helpful. If case (iii) is modified by assuming the $F$ is Lipschitz continuous, then we revert to the Lions-Stampacchia case (in finite dimensions). ⊕

### 2.4.2 Fully Worked Example: Nash Equilibria in m-Person Games

As an example for the use of the theory developed so far for finite dimensional VIs, we may consider the problem of existence of Nash equilibria in $m$-person games, see Sect. 1.2.2, which were connected to VIs in terms of Proposition 1.19.

**Proposition 2.32** *Consider a game satisfying the following:*
*Either*

*(i) All utility functions $x_i \mapsto U_i(x_i; x_{-i})$ are concave and continuously differentiable*
*(ii) All strategy sets $K_i$ are closed, convex, and compact*

*or*

*(i) All utility functions $x_i \mapsto U_i(x_i; x_{-i})$ are strongly concave*[8] *and continuously differentiable*
*(ii) All strategy sets $K_i$ are closed and convex*

*Then this game has a Nash equilibrium. In the second case the Nash equilibrium is unique.*

***Proof*** Under the concavity assumption we have seen (see Proposition 1.19) that $x^* \in C = \prod_{j=1}^m K_j$ is a Nash equilibrium if and only if it solves the VI

$$\langle F(x^*), x - x^* \rangle_{\mathbb{R}^d} \geq 0, \quad \forall\, x \in C, \tag{2.26}$$

where

$$F(x) = (-D_{x_1} U_1, \cdots, -D_{x_m} U_m) : \mathbb{R}^d \to \mathbb{R}^d, \;\; d = \sum_{i=1}^m d_i, \;\; C = \prod_{j=1}^m K_j.$$

Now consider the first set of assumptions. The $C^1$ property of the utility functions guarantees the continuity of $F$. Since $C$ is compact, applying Theorem 2.30(i) yields the existence of a solution to the VI (2.26) and hence of a Nash equilibrium for the game.

[8] Meaning that $-U_i$ are strongly convex, see Example 2.22, 2.

Under the second set of assumptions, the strong concavity of $x_i \mapsto U_i$ implies that $V_i := -U_i$ strongly convex; hence there exists $\rho > 0$ such that, for all $i = 1, \cdots, m$,

$$\langle D_{x_i} V_i(x_i; x_{-i}), z_i - x_i \rangle \geq V_i(z_i; x_{-i}) - V_i(x_i; x_{-i}) - \frac{\rho}{2}\|x_i - z_i\|^2, \quad \forall\, x_i, z_i \in K_i.$$

Swapping the role of $x_i, z_i$ and adding the resulting inequalities, we see that

$$\langle D_{x_i} V_i(z_i; x_{-i}) - D_{x_i} V_i(x_i; x_{-i}), z_i - x_i \rangle \geq \rho \|x_i - z_i\|^2, \quad \forall\, x_i, z_i \in K_i, \quad i = 1, \cdots, m.$$

This result leads to the strong monotonicity of the mapping $F : \mathbb{R}^d \to \mathbb{R}^d$. Continuity follows by the $C^1$ properties of the utility functions. Then applying Theorem 2.30(iii) yields the existence of a unique solution to the VI (2.26) and hence of a Nash equilibrium for the game. □

**Example 2.33** Within the context of Example 1.21, concerning the Cournot game, one can easily check that the matrix $P_0$ is positive definite for $\beta > 0$; hence the mapping $F$ is strongly monotone and Lipschitz continuous (equiv. the utility functions are strongly concave), and hence, there exists a unique Nash equilibrium. ◁

The above existence proof for Nash equilibria can be considered as an alternative to the standard proofs of existence using fixed point arguments.

### 2.4.3 *Fully Worked Example: Existence of Equilibria in Transportation Networks*

Another useful application of the theory developed in this section is in the topic of equilibrium in transportation networks, introduced in Sect. 1.2.1, which according to Proposition 1.13 can be studied in terms of finite dimensional VIs.

**Proposition 2.34** *Assume that the cost function c is strongly monotone and continuous. Then there exists a unique Wardrop equilibrium.*

***Proof*** Recall the equivalence of the Wardrop equilibrium with the VI (1.22). Under the assumption on $c$, we see that $K$ is a strongly monotone operator, so that Theorem 2.30(iii) yields the existence of a unique solution to the VI (1.22) and hence of a Wardrop equilibrium for the game. □

**Example 2.35** Within the context of Example 1.14, consider the transportation network problem with a linear cost function $c(y) = My + c_0$ such that $K_0 = A^T M A$ is strictly positive definite. Then, there exists a unique Wardrop equilibrium. ◁

## 2.5 Infinite Dimensional Theory: Preliminaries

Many problems of interest in management science and finance reduce to variational inequalities which cannot be interpreted as finite dimensional problems.

### *2.5.1 Setting the Scene*

The main object of this section is the theory behind variational inequalities of the form:

**Problem 2.36** Given a convex and closed $C \subset X$ and a mapping $F : X \to Y$ (where $X, Y$ are infinite dimensional Hilbert spaces), find $x^* \in C$ such that

$$\langle F(x^*), x - x^* \rangle \geq 0, \quad \forall\, x \in C, \tag{2.27}$$

where $\langle \cdot, \cdot \rangle$ is a linear pairing operation between the spaces $X$ and $Y$.

In the above, the choice $Y = X'$ is a common choice, whereas if $Y = X$, the $\langle \cdot, \cdot \rangle$ is the inner product in $X$.

**Example 2.37** A prime example of Problem 2.36 is the problem of pricing American or real options. Our starting point will be Proposition 1.25 in Sect. 1.2.3, where the latter problems were reduced to a VI of the form (2.27). To see the connection, let us focus, for simplicity, on the infinite horizon case for one underlying ($N = 1$) and the special case where $D = I = [0, 1]$,

$$\begin{aligned} &\text{Find } u \in K : \int_D (Au - f)(v - u)dx \geq 0, \quad \forall\ v \in K, \\ &K = \{v \ : \ v(x) \geq \psi(x)\ \forall x \in D = [0, 1],\ v(0) = v(1) = 0\}. \end{aligned} \tag{2.28}$$

The unknown quantity here is the value function, $x \mapsto u(x)$, $x \in D$, hence an infinite dimensional object considered as an element of a Hilbert space $X$, with elements $x^* = u$, $x = v$, i.e., functions, and then $K \subset X$ plays the role of $C$ (a closed convex subset of $X$). We also define the mapping $F$ in terms of $v \mapsto F(v) := Av = -u_{xx} + k^2 u$ for every $v \in X$. This is clearly a mapping from the function space $X$ to another function space $Y$ which does not necessarily coincide with $X$. This is easy to see as the operator $A$ may induce the loss of continuous derivatives;[9] if $v \in C^m$, then $Av \in C^{m-2}$. This necessitates the need to define $F : X \to Y$, where $Y$ is a suitable function space in such cases. To cover this need, we require the technical tool of Sobolev spaces, but to make this need clear, assume that

$$X_0 = \{u : [0, 1] \to \mathbb{R} \ : \ u(x) = \sum_n a_n \sin(2\pi n x), \text{ where } \sum_n |a_n|^2 < \infty\}.$$

Readers familiar with Fourier series will realize that $X_0 \subset L^2(0, 1)$ and is a Hilbert space. Formally applying the first derivative on any $u \in X_0$, we see that $u_x = -\sum_n 2\pi n a_n \sin(2\pi n x)$, which makes sense as an element of $L^2(0, 1)$ as long as $\sum_n n^2 |a_n|^2 < \infty$, and hence $u_s \notin X_0$. The situation gets worse for $Au$, where a similar formal calculation will show that

$$(Au)(x) = \sum_n \left((2\pi)^2 n^2 + k^2\right) \sin(2\pi n x), \tag{2.29}$$

so that $Au \in L^2(0, 1)$ if $\sum_n n^4 |a_n|^2 < \infty$. If we chose to define $A$ as a mapping acting on $u \in X_0$, then clearly $Au \notin X_0$. Hence, it is not possible to define $A : X \to X$, for the choice $X = X_0$, but we need a more general definition as $A : X \to Y$ for a suitable $X$ and $Y$, whereas this example shows we may have to choose $X \subset X_0$. One way out would be to consider the spaces $X_m$ as

$$X_m = \{u : [0, 1] \to \mathbb{R} \ : \ u(x) = \sum_n a_n \sin(2\pi n x), \text{ where } \sum_n n^{2m} |a_n|^2 < \infty\} \subset X_0,$$

where $m \in \mathbb{Z}$. It is easy to see that $X_{m+1} \subset X_m$. The space $X_m$, $m \in \mathbb{N}$, consists of these functions in $X_0$ whose $m$-th order derivatives make sense as a Fourier series in $L^2(0, 1)$. Note that the spaces $X_m$ can be understood as sequence spaces in the context of Example 2.12.

In this setting we could then define $A : X_2 \to X_0$. However, as we will see shortly, by a clever construction, we may do much better and define $A$ on a larger space, in terms of $A : X_1 \to X_{-1}$ (where $X_{-1} \supset X_0$ and in fact $X_{-1} = X_1'$, see Example 2.12, setting $w_n = n^2$). Finally the pairing $\langle u, v\rangle := \int_D u(x)v(x)dx$, a quantity related to the inner product in $L^2(D)$, is used. ◁

In order to avoid unnecessary abstraction, we will restrict our analysis of infinite dimensional problems in settings related to that of Example 2.37, so we will always consider problem (2.27) in the context of function spaces with the unknown element being a function $u : D \to \mathbb{R}$.

[9] $v \in C^m$ means that $v$ has a continuous derivative up to order $m$.

**Notation 2.38** *We will from now on adopt the following notation for infinite dimensional problems:*

- $X = \{u : D \subset \mathbb{R}^N \to \mathbb{R}\}$ *a function space. We will no longer use the notation* $x$ *for elements of* $X$ *but replace it by* $u$ *or* $v$ *to signify that this corresponds to a function.*
- $x$ *will still be used but no longer signifies an element of* $X$*. In this setting* $x \in D$ *in the definition of the function* $u \in X$ *as* $x \mapsto u(x)$*.*
- *The notation* $F$ *will be replaced by* $F(u) = A(u) - f$*, where* $A$ *is the operator defined by* $u \mapsto A(u)$ *or simply* $Au$ *if* $A$ *is linear.*
- *The notation* $C$ *for the convex set may at points be replaced by* $K$*, to emphasize that* $K \subset X$*, with* $X$ *a function space.*

We will then specify our study to problems of the form:

**Problem 2.39** Let $X, Y$ be function spaces containing real-valued functions $x \in D \mapsto u(x)$, $C = K \subset X$ closed and convex, and $A : X \to Y$ a mapping.

$$\text{Find } u \in K : \quad \langle A(u), v - u \rangle \geq 0, \quad \forall\, v \in K, \tag{2.30}$$

where $\langle u, v \rangle := \int_D u(x)v(x)dx$ is a bilinear pairing between $X$ and $Y$.

The infinite dimensional nature of problem (2.30) introduces several challenging problems including, the proper formulation of the mapping $A$, i.e., finding the right choice of function spaces $X$ and $Y$, the loss of compactness of closed and bounded sets in infinite dimensional spaces, etc. These intricacies call for the introduction of new tools, such as weak convergence (see Definition 2.7 and Proposition 2.9) or Sobolev spaces (to be introduced shortly in Sect. 2.5.3).

### 2.5.2 The Weak Formulation: Motivation

We now return to the important problem of how to appropriately define the mapping $F(u) = Au$. This calls for a new formulation of the VI, the weak formulation that will be briefly motivated here, before turning to its deeper study later.

Most of the problems to be considered in applications are centered on the operator

$$Au = -\sum_{i=1}^{N} \frac{\partial^2 u}{\partial x_i^2} + k^2 u, \quad k^2 > 0, \tag{2.31}$$

or suitable generalizations (as, e.g., problems related to American option valuation, see Sect. 1.2.3). For such problems it is useful to rewrite (2.30) in a different formulation, called the *weak formulation*. The key observation there is that in its current formulation (2.30) assumes for solutions $u \in C^2(D)$, such that the equivalence with its linear complementarity formulation (see Proposition 1.25) holds. However, the integral defining $\langle A(u), v - u\rangle$ does not require $u \in C^2(D)$, which may not even be feasible, as the simple explicit solution in Example 1.26 indicates.

To clarify this situation, consider for simplicity the case $N = 1$ (corresponding to the single asset case in American options), and assume also that $v, u$ satisfy homogeneous Dirichlet boundary conditions and that $f = 0$. Then, (2.30) by a simple integration by parts argument yields

$$\begin{aligned} 0 \le \langle A(u), v - u\rangle &= \int_D (-\frac{\partial^2 u}{\partial x^2} + k^2 u)(v - u)dx \\ &= \int_D \left(\frac{\partial u}{\partial x}\frac{\partial (v-u)}{\partial x} + k^2 u(v-u)\right)dx, \end{aligned} \tag{2.32}$$

which no longer requires $u \in C^2(D)$ but only $u, v \in C^1(D)$ instead, even allowing for point discontinuities since everything is defined under an integral sign. In fact, using the Lebesgue integral instead of the Riemann integral, we may even relax the assumption $u, v \in C^1(D)$, to the weaker one, $\frac{\partial v}{\partial x}, \frac{\partial v}{\partial x} \in L^2(D)$. The first condition requires continuous derivatives for the functions $u, v$, whereas the second requires only square integrable derivatives for the same functions (note that for such a choice, the integral defining the pairing is well defined). This observation opens up the stage for requiring solutions of VIs of the form (2.30), which are less regular functions than $C^2(D)$, as the classical formulation of the problem requires. Hence upon defining the mapping

$$(u, v) \mapsto \mathbf{a}(u, v) := \int_D \left(\frac{\partial u}{\partial x}\frac{\partial v}{\partial x} + k^2 uv\right)dx = \left\langle \frac{\partial u}{\partial x}, \frac{\partial v}{\partial x}\right\rangle + \langle k^2 u, v\rangle,$$

we may replace (2.30) by the alternative formulation

$$\text{Find } u \in K \ : \ \mathbf{a}(u, v - u) \ge 0, \ \ \forall\, v \in K, \tag{2.33}$$

where now $\langle A(u), v - u\rangle$ with $A : X \to Y$ is replaced by $\mathbf{a}(u, v - u)$, for a mapping $\mathbf{a} : X \times X \to \mathbb{R}$, a move that may allow us to look for solutions to (2.30) in spaces $X$ which are larger than $C^2(D)$ and make life a lot easier. Formulation (2.33), properly defined, will be called the weak formulation of (2.30) as opposed to the original formulation (2.30) or the equivalent LCP, which is called the classical formulation. The above derivation can be easily extended to the multidimensional case. What is missing is making this construction rigorous, i.e., (a) justify all the steps in the above and (b) specify the right setting for the space $X$. Concerning the first task, note that the integration by parts argument in (2.32) works fine if $v \in C_c^1(D)$, a smooth function that is localized in the interior of $D$, so that all integrals are well defined and boundary terms vanish. Concerning the second task, clearly, $v$ is considered as a test function in our variational formulation; hence, we would like to be able to test the VI against a set of possible test functions which is large enough, certainly larger than $C_c^1(D)$. For that, the linearity of $\mathbf{a}$ in the second argument comes to the rescue. We can take as test functions sequences of test functions in $C_c^1(D)$ (so that (2.32) makes sense) and then define $\mathbf{a}(u, v) = \lim_n \mathbf{a}(u, v_n)$, for any $v$ that can be achieved as $v = \lim_n v_n$ for some sequence $(v_n)_{n\in\mathbb{N}} \subset C_c^1(D)$ (clearly employing an appropriate sense of limit, related to an appropriate choice of norm in our function space). This construction makes sense if we choose as $X$ the closure of $C_c^1(D)$, for this appropriate norm.

**Example 2.40** We now revisit Example 2.37 in the above light, using the same notation. We also define the functions $s_n$ by $s_n(x) = \sin(2\pi n x)$. For any $u \in X_2$ and $v \in X_0$, using (2.29) and the orthogonality of the functions $s_n$ in $L^2(0, 1)$, we see that

$$\langle Au, v\rangle = \sum_n ((2\pi)^2 n^2 + k^2) a_n b_n, \tag{2.34}$$

which by the Cauchy-Schwarz inequality for sums is well defined for any $u \in X_2$ and $v \in X_0$.

However, we may rearrange (2.34) as

$$\sum_n ((2\pi)^2 n^2 + k^2) a_n b_n = \sum_n (-2\pi n a_n)(-2\pi n b_n) + \sum_n k^2 a_n b_n,$$

with the RHS making sense as long as both $u, v \in X_1$. Note that (using similar arguments as those that led to (2.34))

$$\mathbf{a}(u, v) = \langle u_x, v_x\rangle + k^2 \langle u, v\rangle = \sum_n (-2\pi n a_n)(-2\pi n b_n) + \sum_n k^2 a_n b_n. \tag{2.35}$$

Note that the series involved in the definition (2.35) are convergent for any $u, v \in X_1$. So essentially instead of defining $A : X_2 \to X_0$, we "extend" it to a larger space, by allowing it to act on functions $u \in X_1$, through the definition of $Au$ on any $u = \sum_n a_n s_n \in X_1$ in terms of

$$\langle Au, v\rangle = \mathbf{a}(u, v) = \sum_n (-2\pi n a_n)(-2\pi n b_n) + \sum_n k^2 a_n b_n, \ \forall\ v = \sum_n b_n s_n \in X_1. \tag{2.36}$$

On the other hand, if we look at $Au = \sum_n ((2\pi)^2 n^2 + k^2) a_n s_n$ itself, for some $u \in X_1$, then clearly $Au \notin X_0$ anymore, since $\sum_n n^2 |a_n|^2 < \infty$ does not necessarily imply that $\sum_n n^4|a_n|^2 < \infty$, but we can easily verify that $Au \in X_{-1}$ since that would require $\sum_n n^{-2}|n^2 a_n|^2 = \sum_n n^2|a_n|^2 \leq \infty$, which is fine for any $u \in X_1$. Hence, in definition (2.36) we may consider $A$ as a mapping $A : X_1 \to X_{-1} = X_1'$.

The weak formulation (2.33) of the VI can then be expressed in terms of the bilinear form

$$\mathbf{a} : X_1 \times X_1 \to \mathbb{R}, \quad \mathbf{a}(u, v) = \sum_n (-2\pi n a_n)(-2\pi n b_n) + \sum_n k^2 a_n b_n,$$

$$\forall\, u = \sum_n a_n s_n, \quad v = \sum_n b_n s_n \in X_1.$$

If $K = E$, a closed subspace of $X_1$, then the above reduces to an equation in terms of the unknown Fourier coefficients $a_n$. ◁

## 2.5.3 *Interlude: Sobolev Spaces*

We need to introduce the concept of the weak derivative:

**Definition 2.41 (Weak Derivatives)** Consider a function $u : D \subset \mathbb{R}^N \to \mathbb{R}$. The function $g$ is called a weak derivative of $u$ with respect to the variable $x_i$, $i = 1, \cdots N$, if

$$\int_D u\phi_{x_i} dx = -\int_D g\phi dx, \quad \forall\, \phi \in C_c^1(D).$$

The function $g$ is denoted as $\frac{\partial u}{\partial x_i} = u_{x_i}$, $i = 1, \cdots, N$, and it is unique (a.e).

The weak derivative allows us to generalize the concept of the derivative to functions that are not differentiable in the classical sense (see Example 2.44). It also allows us to extend the function space $C^1(D)$ to a larger class of vector spaces that may be turned into a Hilbert space with the adoption of the right norm and inner product. These are called Sobolev spaces and play an important role in the study of infinite dimensional VIs of the form (2.30).

Sobolev spaces are function spaces consisting of functions whose weak derivatives are Lebesgue integrable.

**Definition 2.42 (Lebesgue Spaces)** The Lebesgue space $L^p(D)$, where $D \subset \mathbb{R}^N$ open, is defined as

$$L^p(D) = \{u : D \to \mathbb{R}, \;:\; u \text{ Lebesgue measurable}, \quad \int_D |u|^p dx < \infty\},$$

where the integral is considered in the Lebesgue sense.

The Lebesgue spaces $L^p(D)$ are Banach spaces when equipped with the norms

$$\|u\|_{L^p(D)} = \begin{cases} \left(\int_D |u|^p dx\right)^{1/p}, & p \in [1,\infty), \\ \text{ess sup}_{x\in D} |u(x)|, & p = \infty. \end{cases}$$

The case $p = 2$, corresponding to $L^2(D)$, is a Hilbert space endowed with the inner product $\langle u, v\rangle_{L^2(D)} = \int_D uv dx$.

We may now define Sobolev spaces.[10]

**Definition 2.43 (Sobolev Spaces)** The Sobolev space $W^{1,p}(D)$ is defined as

$$W^{1,p}(D) = \Big\{ u \in L^p(D) : \exists\, g_1, \cdots, g_N \in L^p(D),\ s.t.$$

$$\int_D u \frac{\partial \phi}{\partial x_i} dx = -\int_D g_i \phi dx,\ \forall \phi \in C_c^1(D),\ i = 1, \cdots, N \Big\},$$

where the functions $g_i$ are called the weak partial derivatives of $u$ and denoted by $g_i = \frac{\partial u}{\partial x_i} = u_{x_i}$.

The vector $Du = (\frac{\partial u}{\partial x_1}, \cdots, \frac{\partial u}{\partial x_N})$ is the weak analogue of the gradient.

The Sobolev spaces $W^{1,p}(D)$ are Banach spaces when equipped with the norms

$$\|u\|_{W^{1,p}(D)} = \left( \|u\|^p_{L^p(D)} + \sum_{i=1}^N \left\| \frac{\partial u}{\partial x_i} \right\|^p_{L^p(D)} \right)^{1/p}, \quad p \in [1,\infty),$$

reflexive for $p \in (1,\infty)$ and separable for $p \in [1,\infty)$. The case $p = \infty$ is also useful. The case $p = 2$, corresponding to the Sobolev space $W^{1,2}(D)$, is of particular importance as it is a Hilbert space when endowed with the inner product

$$\langle u, v\rangle_{W^{1,2}(D)} = \langle u, v\rangle_{L^2(D)} + \sum_{i=1}^N \langle \frac{\partial u}{\partial x_i}, \frac{\partial v}{\partial x_i} \rangle_{L^2(D)}.$$

**Example 2.44**

1. The function $u(x) = |x|$ defined on $I = (-1, 1)$ does not possess a classical derivative at $x = 0$ but has the weak derivative $u(x) = -\mathbf{1}_{(-1,0)}(x) + \mathbf{1}_{(0,1)}(x)$, as one can see easily by direct application of Definition 2.41. Hence $u \in W^{1,2}(D)$.

[10] Note that alternative (but equivalent) definitions of Sobolev spaces using smoother $\phi$ are encountered.

2. If $u \in W^{1,p}(D)$, then $u^+, u^- \in W^{1,p}(D)$ and $Du^+ = Du\mathbf{1}_{\{u>0\}}$, $Du^- = -Du\mathbf{1}_{\{u<0\}}$.
3. A particular example of Sobolev spaces, in which an explicit construction is possible in terms of Fourier series, is the space $X_1$ of Example 2.37 (see also Example 2.40 in connection to the weak formulation of the operator $A$).

◁

Functions in Sobolev spaces can be understood as limits of sequences of smoother functions in the above norms. As an example we state the following:

**Theorem 2.45 (Meyers-Serrin)** *Let $D \subset \mathbb{R}^N$ be open. Then $C^\infty(D) \cap W^{1,p}(D)$ is dense in $W^{1,p}(D)$, i.e., $W^{1,p}(D)$ can be defined as the completion of $C^\infty(D)$ in the norm $\|\cdot\|_{W^{1,p}(D)}$.*

A very useful Sobolev space is also the space $W_0^{1,p}(D)$. There are two equivalent ways to define this space:

**Definition 2.46 (The Sobolev Spaces $W_0^{1,p}(D)$)**

$$\begin{aligned} W_0^{1,p}(D) &= \{\text{the closure of } C_c^1(D) \text{ in } W^{1,p}(D)\} \\ &= \{u \in W^{1,p}(D) \ : \ u = 0 \text{ on } \partial D\}. \end{aligned}$$

The first definition essentially tells us that the elements in $W_0^{1,p}(D)$ are limits (in the $W^{1,p}(D)$ norm) of sequences of $C_c^1(D)$ functions, i.e., of smooth functions that vanish on the boundary of $D$, denoted by $\partial D$. This information should pass somehow to the limit, i.e., the limit of such a sequence should also vanish at $\partial D$, hence leading to a function whose values on $\partial D$ are equal to 0. However, some care should be taken with this statement. Unlike the case in dimension 1 where Sobolev functions have continuous representatives, so that it makes sense to define their behaviour on the boundary, in higher dimensions the statement $u = 0$ on $\partial D$ should be treated in terms of the trace map $\gamma : W^{1,p}(D) \to L^2(\partial D)$.

The Sobolev spaces $W_0^{1,p}(D)$ are clearly Banach spaces with the norm $\|\cdot\|_{W^{1,p}(D)}$ as subspaces of the original spaces $W^{1,p}(D)$. However, a fundamental result called the Poincaré inequality allows us to define a very convenient equivalent norm for $W_0^{1,p}(D)$.

**Theorem 2.47 (Poincaré Inequality)** *Let $D \subset \mathbb{R}^N$ be bounded. For any $u \in W_0^{1,p}(D)$, there exists a constant $c_P = c_P(D, n)$ such that*

$$\|u\|_{L^p(D)} \leq c_P \|Du\|_{L^p(D)}. \tag{2.37}$$

*Inequality* (2.37)*, which is called the Poincaré inequality, leads to the conclusion that*

$$\|u\|_* := \|Du\|_{L^p(D)} \tag{2.38}$$

*is an equivalent norm for* $W_0^{1,p}(D)$.

Sobolev functions are Lebesgue functions. Sobolev spaces inherit the structure of Lebesgue spaces, as each Sobolev function can be considered as an element of an appropriate Lebesgue space, i.e., Sobolev spaces can be embedded (injected) into suitable Lebesgue spaces or even spaces of continuous functions.

**Theorem 2.48 (Sobolev Embedding)** *Let* $D \subset \mathbb{R}^N$ *be an open set of class* $C^1$ *with* $\partial D$ *bounded, or* $D = \mathbb{R}^N_+$.

(i) $p < N$:

$$\begin{aligned} &W^{1,p}(D) \subset L^{\frac{pN}{N-p}}(D) \text{ continuous injection.} \\ &W^{1,p}(D) \subset L^q(D),\ \forall q \in [1, \frac{pN}{N-p}),\ \text{compact injection if } D \text{ bounded.} \end{aligned} \tag{2.39}$$

(ii) $p = N$:

$$\begin{aligned} &W^{1,p}(D) \subset L^q(D),\ \ \forall q \in [p, \infty) \text{ continuous injection.} \\ &W^{1,p}(D) \subset L^q(D),\ \ \forall q \in [p, \infty),\ \text{compact injection if } D \text{ bounded.} \end{aligned} \tag{2.40}$$

(iii) $p > N$:

$$\begin{aligned} &W^{1,p}(D) \subset L^\infty(D) \text{ continuous injection.} \\ &W^{1,p}(D) \subset C(\bar{D}) \text{ compact injection if } D \text{ bounded.} \end{aligned} \tag{2.41}$$

**Do Remember 2.49 (Convergence in Lebesgue and Sobolev Spaces)**

- $u_n \to u$ *in* $L^p(D)$ *if* $\|u_n - u\|_{L^p(D)} \to 0$ *as* $n \to \infty$. *If* $u_n \to u$ *in* $L^p(D)$, *then there exists a subsequence of* $(u_n)_{n\in\mathbb{N}}$ *such that* $u_n \to u$ *a.e. in D.*
- $u_n \rightharpoonup u$ *in* $L^p(D)$ *if* $\int_D vu_n dx \to \int_D vudx,\ \ \forall v \in L^q(D)$, $p \in (1, \infty)$, *and* $q$ *such that* $\frac{1}{p} + \frac{1}{q} = 1$. *If* $\|u_n\|_{L^p(D)} < c$, *with* $c$ *independent of* $n$,

(continued)

**Do Remember 2.49** (continued)
*then there exists a weakly converging subsequence of* $(u_n)_{n\in\mathbb{N}}$ *in* $L^p(D)$, $p \in (1, \infty)$.

- $u_n \to u$ *in* $W^{1,p}(D)$ *if* $\|u_n - u\|_{W^{1,p}(D)} \to 0$ *as* $n \to \infty$. *In this case* $u_n \to u$ *and* $Du_n \to Du$ *in* $L^p(D)$.
- $u_n \rightharpoonup u$ *in* $W^{1,p}(D)$ *if* $u_n \rightharpoonup u$ *and* $Du_n \rightharpoonup Du$ *in* $L^p(D)$. *If* $\|u_n\|_{W^{1,p}(D)} < c$ *for* $c$ *independent of* $n$, *then there exists a weakly converging subsequence of* $(u_n)_{n\in\mathbb{N}}$ *in* $W^{1,p}(D)$, $p \in (1, \infty)$.
- *If* $W^{1,p}(D) \subset L^s(D)$, *for* $q$ *as in Theorem 2.48, with compact injection, then out of a weakly converging sequence in* $W^{1,p}(D)$ *we can extract a further subsequence strongly converging in* $L^q(D)$.
- *The norms are lower semicontinuous with respect to weak convergence:*

$$u_n \rightharpoonup u \text{ in } L^p(D) \implies \|u\|_{L^p(D)} \le \liminf_n \|u_n\|_{L^p(D)},$$

$$u_n \rightharpoonup u \text{ in } W^{1,p}(D) \implies \|u\|_{W^{1,p}(D)} \le \liminf_n \|u_n\|_{W^{1,p}(D)}.$$

We now briefly discuss the concept of the trace map which provides mathematical rigor to the boundary behaviour of Sobolev functions, which may not be continuous.

**Theorem 2.50 (Traces)** *Let* $D \subset \mathbb{R}^N$, $N \ge 2$, *be bounded and open, with Lipschitz boundary* $\partial D$, *or* $D = \mathbb{R}^{N-1} \times \mathbb{R}_+$. *Then, there exists a unique continuous linear map called trace such that*

$$\gamma : W^{1,2}(D) \to L^2(\partial D) \;:\; \gamma(u) = u|_{\partial D} \;\; \forall u \in W^{1,2}(D) \to L^2(\partial D) \cap C(\bar{D}).$$

*The function* $\gamma(u)$ *is the function prescribing the boundary behaviour of* $u$ *on* $\partial D$ *and is called its trace.*

The functions on $\mathbb{R}^{N-1}$ or $\partial D$ that can be expressed as traces of Sobolev functions in $\mathbb{R}^{N-1} \times \mathbb{R}_+$ or $D$, respectively, form a Banach space, denoted by $H^{1/2}(\partial D)$.

**Definition 2.51 (The Space $H^{1/2}(\partial D)$)** We define

$$H^{1/2}(\partial D) = \gamma(W^{1,2}(D)). \tag{2.42}$$

The space $H^{1/2}(\partial D)$ has some very useful properties that are collected in the following theorem (see, e.g., Theorem 7.52 in Cioranescu et al (2018))

**Theorem 2.52** *Let $D$ be either $\mathbb{R}^{N-1} \times \mathbb{R}_+$ with $\partial D = \mathbb{R}^{N-1}$ or a bounded domain in $\mathbb{R}^N$ with Lipschitz boundary. Then,*

*(i)* $H^{1/2}(\partial D) := \gamma(W^{1,2}(D))$ *is a Banach space when endowed with the norm*

$$\|u\|^2_{H^{1/2}(\partial D)} = \int_{\partial D} |u|(x)^2 dx + \int_{\partial D} \int_{\partial D} \frac{|u(x) - u(y)|^2}{|x - y|^{n+1}} d\mathfrak{S}(x) d\mathfrak{S}(y),$$

*where $\mathfrak{S}$ is the surface measure on $\partial D$ (in the case where $D = \mathbb{R}^{N-1} \times \mathbb{R}_+$, this is the Lebesgue measure on $\mathbb{R}^{N-1}$).*

*(ii) The mapping $\gamma : W^{1,2}(D) \to H^{1/2}(\partial D)$ is linear and continuous.*

*(iii) There exists a linear continuous map, called the lifting operator,*

$$\mathscr{L} : H^{1/2}(\partial D) \to W^{1,2}(D), \quad \textit{defined by}$$

$$g \in H^{1/2}(D) \mapsto \mathscr{L}(g) = \bar{g}, \quad \gamma(\bar{g}) = g, \quad \forall\, g \in H^{1/2}(\partial D).$$

*(iv) The set $\{u|_{\partial D} \; : \; u \in C^\infty(\mathbb{R}^N)\}$ is dense in $H^{1/2}(\partial D)$.*

The lifting operator $\mathscr{L}$ is a very important tool when dealing with nonhomogeneous Dirichlet boundary value problems in dimensions higher than 1. It allows us to take a function $u \in W^{1,2}(D)$ with $u = g$ on $\partial D$ (in the sense of trace) and transform it to $\bar{u} = u - \bar{g} = u - \mathscr{L}(g) \in W_0^{1,2}(D)$. Then, in terms of the above transformation, we can treat a non-homogeneous Dirichlet boundary value problem as a homogeneous boundary value problem. A trivial example of $\mathscr{L}$ is in the case where $D = (0, 1)$, and $g : \partial D = \{0, 1\} \to \mathbb{R}$ is defined by $g(0) = 0$ and $g(1) = 1$. Then, $\mathscr{L}g = u$, where $u : D \to \mathbb{R}$, with $u(x) = x$.

In terms of the trace map, we can define $W_0^{1,2}(D)$ as the space of functions that satisfy $\gamma(u) = 0$.

Sobolev spaces can be extended for the case of higher derivatives, in terms of the following iterative procedure:

**Definition 2.53 (Higher Order Sobolev Spaces)**

$$W^{m,p}(D) := \{u\cdot : D \to \mathbb{R} \; : \; Du \in W^{m-1,p}(D)\}, \quad p, m \in \mathbb{N},$$

where we define $W^{0,p}(D) := L^p(D)$.

A simple example of a higher order Sobolev space that can be explicitly constructed could be $X_m$, $m > 1$ in Example 2.40.

**Remark 2.54** Using this definition we may extend the Sobolev embedding theorem for functions having higher weak derivatives. For example, if $u \in W^{2,p}(D)$, then we may apply Theorem 2.48 for the function $v = Du \in W^{1,p}(D)$ and hence obtain embedding results in Lebesgue spaces of higher exponent $p$ or better continuity properties. In particular, if $p < N$, then by Theorem 2.48, $v = Du \in W^{1,p}(D) \subset L^{\frac{pN}{N-p}}(D)$; hence $u \in W^{1,\frac{pN}{N-p}}(D)$. Set $q = \frac{Np}{N-p}$; if $p > \frac{N}{2}$ (so that $q > N$), then we can apply Theorem 2.48(iii) $W^{1,q}(D) \subset C(\bar{D})$ and conclude that $W^{2,p}(D) \subset C(\bar{D})$ if $p > \frac{N}{2}$. This means, e.g., that even though in $N = 2$ a $W^{1,2}(D)$ function is not continuous, a $W^{2,2}(D)$ function will be. With similar arguments, if $u \in W^{2,p}(D)$ for $p > N$, then $Du \in W^{1,p}(D) \subset C(\bar{D})$; hence $u \in C^1(\bar{D})$. ⊕

In what follows, the function spaces $X = W_0^{1,2}(D)$ or $X = W^{1,2}(D)$ will be the basis for the study of the weak solutions for VIs. We will also require higher Sobolev spaces in the study of problems of regularity of solutions.

## 2.6 A General Class of Elliptic (Time Stationary) Variational Inequalities

We now introduce a general class of variational inequalities of the form (2.30) (see Problem 2.39) in higher spatial dimensions and for general operators, of the type usually encountered in optimal stopping problems which appear in option pricing. We will present a general theory for these problems.

### *2.6.1 A Class of Linear Differential Operators, Their Corresponding Bilinear Forms, and Weak Forms of VIs*

We will from now on consider VIs involving linear operators $A$ which are generalizations of (2.31) in the sense that we will now allow for varying coefficients. This is useful because even though we have managed to reduce the optimal stopping problem for the case where the underlying followed the geometric Brownian motion to a constant coefficient problem, we may not always be so lucky with more general models, and it is useful to develop our tools for a more general class of problems.

**Definition 2.55 (Operators, Bilinear Forms, and VIs)** Let $X = W^{1,2}(D)$ or $X = W_0^{1,2}(D)$ and $C = K \subset X$ be a convex and closed set. We consider:

- Operators in the form

$$Au := -\sum_{i=1}^{N} \frac{\partial}{\partial x_i}\left(\sum_{j=1}^{N} a_{ij}\frac{\partial u}{\partial x_j}\right) + \sum_{i=1}^{N} b_i \frac{\partial u}{\partial x_i} + cu, \tag{2.43}$$

  where $\mathscr{A} = (a_{ij})_{i=1,\cdots,N}^{j=1,\cdots,N}$, $b = (b_1, \cdots, b_N)$, and $c$ are given functions of $x$
- The related bilinear form $\mathbf{a} : X \times X \to \mathbb{R}$ defined by

$$\begin{aligned}\mathbf{a}(u, v) &= \int_D \left(\sum_{i=1}^{N}\sum_{j=1}^{N} a_{ij}\frac{\partial u}{\partial x_i}\frac{\partial v}{\partial x_j} + \sum_{i=1}^{N} b_i \frac{\partial u}{\partial x_i} v + cuv\right) dx \\ &= \int_D aDu \cdot Dvdx + \int_D b \cdot Du\, vdx + \int_D cu\, vdx, \quad \forall\, u, v \in X,\end{aligned} \tag{2.44}$$

  where by $Du$ we denote the gradient of the function $u$ and by $\cdot$ the Euclidean inner product in $\mathbb{R}^N$
- The corresponding weak formulation of the VI

$$u \in K \ : \ \mathbf{a}(u, v-u) - \langle f, v-u\rangle = \langle Au - f, v-u\rangle \geq 0, \ \ \forall\, v \in K \tag{2.45}$$

To justify that (2.45) is a weak formulation of a VI, just revisit the arguments of Sect. 2.5.2 keeping in mind the density results of smooth test functions in Sobolev spaces, provided in Sect. 2.5.3.

**Example 2.56** Consider the weak problem

$$\mathbf{a}(u, v-u) - \langle f, v-u\rangle \geq 0, \ \ \forall\, v \in K,$$

$$K = \{v \in W^{1,2}(D), \ : \ v \geq \psi, \ \text{in } D, \ v = g \text{ on } \partial D\},$$

where by $v = g$ on $\partial D$ we mean that $\gamma(v) = g$, where $\gamma$ is the trace map (see Theorem 2.50). In the case where $g = 0$ (homogeneous Dirichlet conditions), we work in the space $W_0^{1,2}(D)$. Note that even in the case where $g \neq 0$, the function $v - u$, which is essentially our test function, has vanishing trace, i.e., $v - u \in W_0^{1,2}(D)$.

This observation is important, as it allows us to eliminate all boundary terms arising in the integration by parts which leads us to the definition of the bilinear form (2.44).

An alternative way to look at this problem is to use the lifting map $\mathscr{L}$ (see Theorem 2.52). If $g \in H^{1/2}(\partial D)$, then there exists $\bar{g} := \mathscr{L}(g) \in W^{1,2}(D)$ such that $\gamma(\bar{g}) = g$. Working in terms of $\bar{u} = u - \bar{g} \in W_0^{1,2}(D)$, we express the VI in the new function $\bar{u}$ and the corresponding test function $\bar{v} = v - \bar{g} \in W_0^{1,2}(D)$ and rewrite the VI as a weak problem using $X = W_0^{1,2}(D)$, with a modified obstacle $\bar{\psi} = \psi - \bar{g}$. This leads to the variational problem

$$\mathbf{a}(\bar{u}, \bar{v} - \bar{u}\rangle - \langle f, \bar{v} - \bar{u}\rangle + \mathbf{a}(\bar{g}, \bar{v} - \bar{u}), \ \forall \bar{v} \in \bar{K},$$

$$\bar{K} = \{\bar{v} \in W_0^{1,2}(D), \ : \ \bar{v} \geq \bar{\psi}, \text{ in } D\}.$$

The choice $X = W^{1,2}(D)$ can also accommodate Neumann boundary conditions. ◁

**Remark 2.57** The class of operators in (2.43) and the corresponding VIs (2.45) include as special cases the problems introduced in Sect. 2.45 in relation to optimal stopping problems and the valuation of American or real options, in the single or multiple asset case. Due to the fact that the coefficients are allowed to be functions of the spatial variable $x$, this formulation may also allow for treatment of more general models (or even of the GBM model without resorting to additional variable changes). While a typical operator arising in Itō calculus is of the slightly different form[11]

$$\bar{A}u := -Tr(\bar{\mathscr{A}} D^2 u) + \bar{b} \cdot Du + \bar{c}u, \tag{2.46}$$

where by $Du$ we denote the gradient of $u$ and by $D^2u$ we denote the Hessian of $u$, this operator under suitable smoothness conditions on the coefficients may be transformed to an operator of the form (2.43)[12] by a suitable modification of the function $b$. ⊕

We now collect some properties of the bilinear form **a** defined in (2.44) that will be required for the well-posedness of the corresponding VI.

---

[11] The first term is related to the fluctuations generated by the Wiener process, and the second is related to the drift.

[12] Note that the operators (2.46) are related to the class of operators (2.43): If we choose $\mathscr{A} = \bar{\mathscr{A}}$, $\bar{b} = (\bar{b}_1, \cdots, \bar{b}_N)$ with $\bar{b}_i = b_i - \sum_{j=1}^{N} \frac{\partial a_{ji}}{\partial x_j}$, $i = 1, \cdots, N$, and $\bar{c} = c$, we have that $\bar{A} = A$.

**Definition 2.58 (Coercive and Continuous Form)** Let $\mathbf{a} : X \times X \to \mathbb{R}$ be a bilinear form on the Hilbert space $X$. The form $\mathbf{a}$ is called:

(i) Coercive if

$$\mathbf{a}(u, u) \geq \mathfrak{c}_0 \|u\|^2, \quad \text{for some } \mathfrak{c}_0 > 0, \quad \forall\, u \in X \tag{2.47}$$

(ii) Continuous if

$$|\mathbf{a}(u, v)| \leq \mathfrak{c}_1 \|u\| \|v\|, \quad \text{for some } \mathfrak{c}_1 > 0, \quad \forall\, u, v \in X \tag{2.48}$$

We will require the following condition on the coefficients $\mathscr{A}, b, c$, so that $\mathbf{a}$ is continuous and coercive:

**Assumption 2.59 (Ellipticity Condition)** *Assume that there exists a constant* $\alpha_0 > 0$ *such that*[13]

$$\mathscr{A}(x)\xi \cdot \xi := \langle \mathscr{A}(x)\xi, \xi \rangle_{\mathbb{R}^N} \geq \alpha_0 |\xi|^2, \quad \forall\, \xi \in \mathbb{R}^N, \quad a.e.\ x \in \mathbb{R}^N. \tag{2.49}$$

*This property is called the ellipticity condition for the operator A.*

The following proposition provides conditions for the bilinear form $\mathbf{a}$ to be continuous and coercive (the conditions are not optimal and can be relaxed).

**Proposition 2.60** *Let the ellipticity condition* (2.49) *of Assumption 2.59 hold. Moreover, assume that* $\inf_{x \in D} c(x) \geq c_0 > 0$ *and that* $b$ *is sufficiently small, for instance,* $\|b\|_{L^\infty(D)} \leq \left(\frac{\alpha_0}{c_0}\right)^{1/2}$. *Then, the operator A in Definition 2.55 (see* (2.44)*) is continuous and coercive in* $W^{1,2}(D)$,

***Proof*** We will first show continuity and the coercivity for $\mathbf{a}$.

1. We recall the Hölder inequality, according to which for any scalar functions $f, g, w$ such that $f \in L^\infty(D)$, $g, w \in L^2(D)$, it holds that

$$\int_D |f\, g\, w| dx \leq \|f\|_{L^\infty(D)} \|g\|_{L^2(D)} \|w\|_{L^2(D)}.$$

[13] Or in coordinate form $\sum_{i,j=1}^N a_{ij}\xi_i\xi_j \geq \alpha_0 \sum_{i=1}^N \xi_i^2, \ \forall\, \xi \in \mathbb{R}^N$.

This can be extended to matrix-valued and vector-valued functions (by properly adapting the definition of the relevant norms of course), so that if $\mathscr{A} \in L^\infty(D)$, then

$$|\int_D \mathscr{A} Du \cdot Dv dx| \le \|\mathscr{A}\|_{L^\infty(D)} \|Du\|_{L^2(D)} \|Dv\|_{L^2(D)}.$$

With that in mind, we have

$$\begin{aligned}
|\mathbf{a}(u, v)| &\le \int_D |\mathscr{A} Du \cdot Dv + b \cdot Duv + cuv| dx \\
&\le c_1 \|Du\|_{L^2(D)} \|Dv\|_{L^2(D)} + c_2 \|Du\|_{L^2(D)} \|v\|_{L^2(D)} + c_3 \|u\|_{L^2(D)} \|v\|_{L^2(D)} \\
&\le c(\|Du\|_{L^2(D)} \|Dv\|_{L^2(D)} + \|u\|_{L^2(D)} \|v\|_{L^2(D)}) \\
&\le c(\|Du\|^2_{L^2(D)} + \|u\|^2_{L^2(D)})^{1/2} (\|Dv\|^2_{L^2(D)} + \|v\|^2_{L^2(D)})^{1/2} \\
&= c\|u\|_{W^{1,2}(D)} \|v\|_{W^{1,2}(D)},
\end{aligned} \tag{2.50}$$

which proves continuity of **a**.

2. To prove coercivity (using the notation $\xi \cdot \zeta = \langle \xi, \zeta \rangle_{\mathbb{R}^N}$),

$$\begin{aligned}
\mathbf{a}(u, u) &= \int_D \mathscr{A} Du \cdot Du dx + \int_D b \cdot Duu dx + \int_D cu^2 dx \\
&\overset{(2.49)}{\ge} \alpha_0 \int_D |Du|^2 dx + \int_D b \cdot Duu dx + \int_D cu^2 dx.
\end{aligned} \tag{2.51}$$

We consider two possible cases:

(i) If $b = 0$ and $c \ge c_0 > 0$, where $c_0$ is a suitable positive constant, then by (2.51)

$$\mathbf{a}(u, u) \ge \alpha_0 \|Du\|^2_{L^2(D)} + c_0 \|u\|^2_{L^2(D)} \ge \min\{\alpha_0, c_0\} \|u\|^2_{W^{1,2}(D)}. \tag{2.52}$$

(ii) If $b \neq 0$ and $c \ge c_0 > 0$, where $c_0$ is a suitable positive constant, then by (2.51)

$$\begin{aligned}
\mathbf{a}(u, u) &\ge \alpha_0 \|Du\|^2_{L^2(D)} - \int_D |(b \cdot Du) u| dx + c_0 \|u\|^2_{L^2(D)} \\
&\ge \alpha_0 \|Du\|^2_{L^2(D)} - \|b\|_{L^\infty(D)} \|Du\|_{L^2(D)} \|u\|_{L^2(D)} + c_0 \|u\|^2_{L^2(D)} \\
&\ge \alpha_0 \|Du\|^2_{L^2(D)} - \|b\|_{L^\infty(D)} \epsilon^2 \|Du\|^2_{L^2(D)} - \frac{1}{4\epsilon^2} \|u\|^2_{L^2(D)} + c_0 \|u\|^2_{L^2(D)} \\
&= \left(\alpha_0 - \|b\|_{L^\infty(D)} \epsilon^2\right) \|Du\|^2_{L^2(D)} + \left(c_0 - \frac{1}{4\epsilon^2}\right) \|u\|^2_{L^2(D)},
\end{aligned} \tag{2.53}$$

where $\epsilon > 0$ is as yet arbitrary. The idea now is to choose $\epsilon$ such that the RHS of (2.53) is larger than some positive multiple of $\|u\|^2_{W^{1,2}(D)}$. For example, choosing

$$\epsilon = \left(\frac{1}{2c_0}\right)^{1/2}, \quad \text{and if} \quad \|b\|_{L^\infty(D)} \leq \left(\frac{\alpha_0}{c_0}\right)^{1/2},$$

then,[14] so that (2.53) yields

$$\mathbf{a}(u,u) \geq \frac{\alpha_0}{2}\|Du\|^2_{L^2(D)} + \frac{c_0}{2}\|u\|^2_{L^2(D)} \geq \frac{1}{2}\min\{\alpha_0, c_0\}\|u\|^2_{W^{1,2}(D)}, \tag{2.54}$$

which is the required coercivity estimate. The proof is complete.

□

**Remark 2.61** Similarly we may obtain continuity and coercivity of $\mathbf{a}$ in $W^{1,2}_0(D)$. In fact the coercivity estimates in $W^{1,2}_0$ may be easier to obtain taking into account the Poincaré inequality (2.37) (see Theorem 2.47) and the fact that $\|Du\|_{L^2(D)}$ is an equivalent norm in $W^{1,2}_0(D)$. ⊕

## 2.6.2 Getting Technical: Variational Definition of the Operator A

Note that we have been deliberately silent[15] about the function space $Y$ where the operator $A : X \to Y$ takes values in, and preferred to work in terms of the bilinear form $\mathbf{a} : X \times X \to \mathbb{R}$ and define the operator $A$ in terms of

$$\langle Au, v\rangle = \int_D (Au)v dx := \mathbf{a}(u,v), \;\; \forall\, u, v \in X, \tag{2.55}$$

instead, where $\langle\cdot,\cdot\rangle = \langle\cdot,\cdot\rangle_{Y,X}$ is the pairing between the spaces $Y$ and $X$; for the particular cases considered here we will focus on the choice $X = W^{1,2}_0(D)$ or $W^{1,2}(D)$. We will show here that a suitable definition for $A$, compatible with (2.55), is $A : X \to X'$.

Indeed, definition (2.55) (called the variational definition) can be interpreted as $Au \in Y$ with $Y$ chosen so that the integral defining the pairing $\langle\cdot,\cdot\rangle = \langle\cdot,\cdot\rangle_{Y,X}$ makes sense for every $v \in X$. Since $X \subset L^2(D)$, the above integration $\langle w, v\rangle :=$

---

[14] With this choice $\left(c_0 - \frac{1}{4\epsilon^2}\right) = \frac{c_0}{2}$, whereas $\left(\alpha_0 - \|b\|_{L^\infty(D)}\epsilon^2\right) = \alpha_0 - \frac{\|b\|_{L^\infty(D)}}{2c_0} \geq \frac{\alpha_0}{2}$.

[15] The reader may at this point wish to refer to Example 2.40.

$\int_D wvdx$ may make sense even if $w$ is an element of a space larger than $L^2(D)$, so that the space $Y$ can be chosen as the larger function space for which the integral $\int_D wvdx$ is well defined for any $v \in X$. Readers familiar with functional analysis will identify $Y$ as the dual space of $X$, denoted by $X'$ (see Sect. 2.2.1). Hence, in the sense of (2.55) we may consider $Au$ as an element of $X'$, i.e., $A : X \to X'$. In the present context $X \subset X_0 := L^2(D) \subset X'$, with continuous injection $X \subset X_0$, a construction called a Gelfand triple, with important applications in the theory of PDEs and VIs. The intermediate space $X_0$ is called the pivot space and plays an important role in our construction. Moreover the injections are dense.

The bilinearity of the form $\mathbf{a}$, along with (2.55), implies that the operator $A : X \to X'$ is linear. Combining the definition of the operator $A : X \to X'$ in (2.55) with the coercivity and continuity of $\mathbf{a}$ yields[16]

$$\begin{aligned} \langle Au, u\rangle_{X',X} &:= \mathbf{a}(u,u) \geq c_0\|u\|_X^2, \quad \forall v \in X, \\ \|Au\|_{X'} &:= \sup_{v\in X} \frac{|\langle Au, v\rangle_{X',X}|}{\|v\|_X} \leq c_1\|u\|_X, \end{aligned} \tag{2.56}$$

corresponding to a monotonicity and a continuity condition for $A$, respectively.

Moreover, by the Riesz representation theorem (see Theorem 2.11), there exists a unique element $i(Au) \in X$ such that

$$\begin{aligned} \mathbf{a}(u,u) &= \langle Au, u\rangle_{X',X} = \langle i(Au), u\rangle_X, \\ \|i(Au)\|_X^2 &= \langle i(Au), i(Au)\rangle_X = \|Au\|_{X'}^2. \end{aligned} \tag{2.57}$$

We emphasize that $Au \in X'$ and $i(Au) \in X$ are in principle different entities.

### 2.6.3 An Existence Result

The assumptions of continuity and coercivity of the bilinear form $\mathbf{a}$ and the corresponding properties of the mapping $F(u) = Au - f$ imply that the VI falls within the domain of application of the abstract Lions-Stampacchia Theorem 2.28. Moreover, the general results given in Sect. 2.3.2, such as, e.g., Minty's reformulation, hold.

[16] For the first estimate recall that

$$|\langle Au, v\rangle_{X',X}| = |\mathbf{a}(u,v)| \overset{(2.48)}{\leq} c_1\|u\|_X\|v\|_X \implies \frac{|\langle Au, v\rangle_{X',X}|}{\|v\|_X} \leq c_1\|u\|_X, \ \forall v \in X.$$

**Theorem 2.62 (Lions-Stampachia)** *Let $C = K$ be a closed convex subset of a Hilbert space $X$ and $\mathbf{a} : X \times X \to \mathbb{R}$ be a continuous and coercive bilinear form (as in Definition 2.55).*

*Then:*

*(i) The variational inequality*

$$u \in K \ : \ \mathbf{a}(u, v-u) \geq \langle f, v-u \rangle, \quad \forall\, v \in K \tag{2.58}$$

*admits a unique solution, for every $f \in X$ (in fact for every $f \in X'$ where by $X'$ we denote the dual space of $X$).*

*(ii) If $\mathbf{a}$ is symmetric, then (2.58) is equivalent to the constrained minimization problem*

$$\min_{u\in C} \frac{1}{2}\mathbf{a}(u,u) - \langle f, u\rangle .$$

*(iii) The solution mapping $S : X \to C = K$ (respectively, $S : X' \to C = K$) defined by $f \mapsto S(f) := u$, where $u$ is the solution of (2.58), is a Lipschitz map satisfying*

$$\|S(f_1) - S(f_2)\| \leq \frac{1}{c_0}\|f_1 - f_2\|, \quad \forall\, f_1, f_2 \in X \tag{2.59}$$

*(with the RHS replaced by $\|f_1 - f_2\|_{X'}$ in the case where we consider the data $f_1, f_2 \in X'$).*

***Proof***

(i) It follows directly from the abstract Lions-Stampacchia Theorem 2.28.

(ii) Define the function $v \mapsto \Phi(v)$ by $\Phi(v) = \frac{1}{2}\mathbf{a}(v,v) - \langle f, v\rangle$. Let $u$ be the solution (2.58). Using symmetry and bilinearity of $\mathbf{a}$, we see that

$$\begin{aligned}\Phi(v) = \Phi(u + (v-u)) &= \Phi(u) + \underbrace{\frac{1}{2}\mathbf{a}(v-u, v-u)}_{\geq 0} + \underbrace{\mathbf{a}(v-u, u) - \langle f, v-u\rangle}_{\geq 0} \\ &\geq \Phi(u),\end{aligned}$$

where we used coercivity for the first underbraced term and symmetry combined with the fact that $u$ solves the VI for the second.

(iii) To show that the solution map is Lipschitz, we essentially need the coercivity property of $\mathbf{a}$. Let $u_i$ be the solution of (2.58) with $f_i$ on the RHS, $i = 1, 2$. We take $v = u_2$ in the VI for $u_1$ and $v = u_1$ in the VI for $u_2$ and add to obtain

$$\mathbf{a}(u_1 - u_2, u_1 - u_2) \leq \langle f_1 - f_2, u_1 - u_2 \rangle. \tag{2.60}$$

We now continue as follows:

$$\begin{aligned} c_0 \|u_1 - u_2\|_X^2 &\overset{(2.47)}{\leq} \mathbf{a}(u_1 - u_2, u_1 - u_2) \overset{(2.60)}{\leq} \langle f_1 - f_2, u_1 - u_2 \rangle \\ &\leq |\langle f_1 - f_2, u_1 - u_2 \rangle| \leq \|f_1 - f_2\| \|u_1 - u_2\|_X, \end{aligned} \tag{2.61}$$

which upon dividing (2.61) by $\|u_1 - u_2\|_X$ leads to the required result. Note that when performing the estimate $|\langle f, w \rangle| \leq \|f\| \|w\|_X$, we deliberately have not specified the norm of $f$: It is to be understood as $\|f\|_{X'}$ if $f \in X'$ (and in this case the estimate follows from the definition of the dual norm) or as $\|f\|_X$ if $f \in H$ (and in this case the estimate follows from the Cauchy-Schwarz inequality).

□

**Remark 2.63** In the case where $C = E$ a closed linear subspace, we recover the celebrated Lax-Milgram lemma (Lax and Milgram (1954)) which is a powerful tool for the solution of equations of the form $Au = f$, through their variational formulation $\langle Au - f, v \rangle = 0$, for all $v \in E$. ⊕

**Do Remember 2.64** *Applying the Lions-Stampacchia theorem for operators of the form given in Definition 2.55 and selecting* $X = W_0^{1,2}(D)$ *or* $X = W^{1,2}(D)$*, we obtain solvability results for VIs of the form* (2.45)*, for Dirichlet or Neumann boundary conditions (which set the derivative in the normal direction on the boundary equal to* 0*).*

### 2.6.4 *Comparison and Maximum Principles*

Comparison principles can be very informative concerning the behaviour of solutions. For example, they may answer interesting questions such as "what would happen to the solution of the VI if the obstacle or the boundary data or even the form of the operator changes?"

Consider the bilinear forms $\mathbf{a}_i : X \times X \to \mathbb{R}$, the closed convex sets $C_i = K_i$, and $f_i \in X, i = 1, 2$. We will provide here a general comparison result for the VIs

$$u_i \in K_i \ : \ \mathbf{a}_i(u_i, v_i - u_i) \geq \langle f_i, v_i - u_i \rangle, \ \ \forall\, v_i \in K_i, \ \ i = 1, 2. \tag{2.62}$$

**Assumption 2.65** *Assume that the data of problem* (2.62) *satisfy the following:*

*(i)* $\langle A_1 v_1 - A_2 v_2, (v_1 - v_2)^+ \rangle \geq 0$, *for every* $v_i \in K_i$, $i = 1, 2$.
*(ii)* *For every* $v_1 \in K_1$, $v_2 \in K_2$, *it holds that* $\max(v_1, v_2) \in K_1$ *and* $\min(v_1, v_2) \in K_2$,
*(iii)* $f_1 \geq f_2$.

The following examples clarify Assumption 2.65.

**Example 2.66** Let $A_i = A_0 + c_i I$, where $A_0$ is a coercive operator, of the form (2.43), with $b = 0$. Then $A_1, A_2$ satisfy Assumption 2.65(i) for $v \geq 0$, if $c_1 \geq c_2 > 0$.

Indeed for $A_i$ as above,

$$\begin{aligned}\langle A_1 v_1 - A_2 v_2, (v_1 - v_2)^+ \rangle &= \langle A_0(v_1 - v_2), (v_1 - v_2)^+ \rangle \\ &\quad + \langle c_1 v_1 - c_2 v_2, (v_1 - v_2)^+ \rangle \\ &= \langle A_0(v_1 - v_2)^+, (v_1 - v_2)^+ \rangle + \langle c_1 v_1 - c_2 v_2, (v_1 - v_2)^+ \rangle \\ &\geq \langle c_1 v_1 - c_2 v_2, (v_1 - v_2)^+ \rangle,\end{aligned}$$

where we used the fact that (a) for any function $v = v^+ - v^-$, it holds that $\langle A_0 v^-, v^+ \rangle = 0$ and (b) $A_0$ is coercive. It remains to estimate the last term

$$\langle c_1 v_1 - c_2 v_2, (v_1 - v_2)^+ \rangle = \langle (c_1 - c_2) v_1, (v_1 - v_2)^+ \rangle + \langle c_2 v_1, (v_1 - v_2)^+ \rangle \geq 0,$$

where we used $c_1 \geq c_2$ and $v_1, v_2 \geq 0$. ◁

**Example 2.67** Let $C_i = K_i = \{v \in W^{1,2}(D) \ : \ v \geq \psi_i, \ v = g_i \text{ on } \partial D\}$. Then $K_1, K_2$ satisfy Assumption 2.65(ii) if $\psi_1 \geq \psi_2$ and $g_1 \geq g_2$.

Indeed, let $v_1 \in K_1$ and $v_2 \in K_2$. Then,

$$\begin{cases} u_1 \geq \psi_1, \\ u_2 \geq \psi_2 \end{cases} \Longrightarrow \max(u_1, u_2) \geq u_1 \geq \psi_1 \Longrightarrow \max(u_1, u_2) \in K_1,$$

while

$$\begin{cases} u_1 \geq \psi_1 \geq \psi_2, \\ \quad u_2 \geq \psi_2 \end{cases} \Longrightarrow \min(u_1, u_2) \geq \psi_2 \Longrightarrow \min(u_1, u_2) \in K_2,$$

with a similar argument following for the boundary data. ◁

We may now obtain the following comparison principle for solutions of VIs.

**Proposition 2.68 (Comparison Principle)** *Under Assumption 2.65 the solutions of the VI* (2.62) *satisfy the comparison relation* $u_1 \geq u_2$.

***Proof*** The strategy of the proof is to use an appropriate test function for each variational inequality. In particular, we shall use $v_1 = \max(u_1, u_2) = u_1 + (u_2 - u_1)^+ \in K_1$ (by Assumption 2.65(ii)) as test function in the VI for $u_1$ and $v_2 = \min(u_1, u_2) = u_2 - (u_2 - u_1)^+ \in K_2$ (by Assumption 2.65(ii)) as test function in the VI for $u_2$. Adding the resulting inequalities and rearranging, we obtain

$$\langle A_2 u_2, -A_1 u_1, (u_2 - u_1)^+ \rangle + \langle f_2 - f_1, (u_2 - u_1)^+ \rangle \leq 0.$$

Then, Assumption 2.65(i) implies that

$$\langle f_2 - f_1, (u_2 - u_1)^+ \rangle = \int_D (f_2 - f_1)(u_2 - u_1)^+ dx \leq 0,$$

which by Assumption 2.65(iii) yields $(u_2 - u_1)^+ = 0$ a.e., hence $u_1 \geq u_1$ a.e. □

## 2.6.5 *Fully Worked Example: Perpetual American Basket Options*

As an example of the applicability of the general theory developed thus far, we consider the problem of valuation for American basket options or real options, over $N$ underlying assets, under the multidimensional GBM model introduced in Sect. 1.2.3. We will assume that $x = (x_1, \cdots, x_N)$ takes values in an open bounded set $D \subset \mathbb{R}^N$, which contains all the possible realistic values that the asset prices may attain. We will denote by $\partial D$ the boundary of $D$ and define a function $g : \partial D \to \mathbb{R}$ that describes the value of the option when the price of the underlying attains their limiting values on $\partial D$. An example for $D$ could be $D = (-L, L)^N$, and the function $g$ may be a constant on $\partial D$ or even vanishing (corresponding to the case of homogeneous Dirichlet boundary conditions, modelling barrier features of the option).

Using Proposition 1.25 the value function is related to the solution of the VI

$$\text{Find } u \in K : \langle Au - f, v - u \rangle := \int_D (Au - f)(v - u)dx \geq 0, \quad \forall\, v \in K,$$
$$K = \{v \ : \ v(x) \geq \psi(x) \ \forall x \in D, \ \ v = g \text{ on } \partial D\}. \tag{2.63}$$

We will consider the case where the operator $A$ is chosen as in (1.47)

$$Au(x) = -\frac{1}{2} \sum_{i,j=1}^{N} a_{ij} \frac{\partial^2 u}{\partial x_i \partial x_j} + k^2 u, \quad k^2 = r + \frac{1}{2} \alpha^T \mathscr{A} \alpha, \tag{2.64}$$

where $x_i = \ln y_i$, with $y_i$ being the current price of asset $i$, and $\psi$ the pay-off function (refer to Sect. 1.2.3 for the modelling). Even though the exact form of the projection operator $\Pi_K$ is not required, it can be obtained explicitly in the form $\Pi_K(u) = (u - \psi)^+ + \psi$ (see Example 2.18). Moreover, $v \geq \psi$ can be interpreted in the a.e. sense.

We will now use the general framework developed in the previous sections to study the existence of a weak solution for VI (2.63), thus providing well-posedness results for the American basket option pricing problem for the multidimensional GBM model. The appropriate function space framework here is the use of the Hilbert space $X = W^{1,2}(D)$ which may accommodate for the non-homogeneous boundary conditions. If $g = 0$, then we may use the setting $X = W_0^{1,2}(D)$, which is in fact simpler.

The following proposition is the main result of this section.

**Proposition 2.69** *If $\mathscr{A}$ is a positive definite matrix, then problem* (2.63) *admits a unique solution $u \in K \subset X = W^{1,2}(D)$ for any $f \in L^2(D)$ and $g \in H^{1/2}(\partial D)$.*

***Proof*** The proof proceeds by direct use of the Lions-Stampacchia theorem (see Theorem 2.62).

We first check that the conditions of this theorem are satisfied: The positivity of $\sigma$ guarantees the ellipticity of the operator $A$ (see Assumption 2.59). Moreover, if the matrix $\mathscr{A}$ is positive definite, then $k^2 > 0$. The operator $A$ is a constant coefficient operator (of the form (2.44) of Definition 2.55 with $b = 0$); hence all the assumptions of Proposition 2.60 are trivially satisfied, and continuity and coercivity of $A$ follows by a straightforward application of this proposition.

We now show that $K$ as above is convex and closed. Convexity is immediate. To show closedness consider any sequence $(u_n)_{n\in\mathbb{N}} \subset K$ such that $u_n \to u$ in $W^{1,2}(D)$. We will show that $u \in K$. To show that, observe that by definition $u_n \to u$ in $W^{1,2}(D)$ implies that $u_n \to u$ in $L^2(D)$ and $Du_n \to Du$ in $L^2(D)$. Since[17] $u_n \to u$ in $L^2(D)$, there exists a further subsequence $(u_{n_k})_{k\in\mathbb{N}}$ such that $u_{n_k} \to u$ a.e. in $D$. Since $u_n \in K$, we have that $u_n \geq \psi$ for every $n$, and working with this subsequence and passing to the limit we get that $u \geq \psi$ a.e. It remains to check the boundary condition. Since $u_n \in K$, for any $n$ it holds that $\gamma(u_n) = g$. The map $\gamma : W^{1,2}(D) \to H^{1/2}(\partial D)$ is continuous, so that passing to the limit in the above we get that $\gamma(u) = g$. Hence $u \in K$ and $K$ is closed.

We note here that when using this method, it is required to assume that $k^2 > 0$ since the bilinear form $\mathbf{a}$ is not coercive in $W^{1,2}(D)$ for $k^2 = 0$ (in contrast to what happens with this form in $W_0^{1,2}(D)$ as a consequence of the Poincaré inequality).

An alternative way to prove the result is to transform it into a homogeneous problem by using the lifting map $\mathscr{L}$, as in Example 2.56, and work using the variational formulation in $X = W_0^{1,2}(D)$. □

[17] We note here that even if $u_n \rightharpoonup u$ in $W^{1,2}(D)$, for $D$ such that the compact embedding $W^{1,2}(D) \subset L^2(D)$ (up to a subsequence), we can still have $u_n \to u$ in $L^2(D)$ (see Do Remember 2.49).

**Remark 2.70** The positivity of $\mathscr{A}$ in this constant coefficient case allows for the transformation of variables $x = \mathfrak{D}z$, where $\mathscr{A} = \mathfrak{D}\mathfrak{D}^T$, which transforms the operator to $\hat{A} = -\frac{1}{2}\Delta_z + k^2 I$, where $\Delta_z$ is the Laplacian (see Sect. 1.2.3). ⊕

Having at hand the existence result we may obtain interesting qualitative results, as, for example, the following comparison principle.

**Proposition 2.71** *Let* $u_i := u_{f_i,\psi_i,g_i}$ *be the solution of* (2.63) *corresponding to the data* $f_i$, $\psi_i$, *and* $g_i$, *for* $i = 1, 2$. *If* $f_1 \geq f_2$, $\psi_1 \geq \psi_2$, *and* $g_1 \geq g_2$, *then* $u_1 \geq u_2$.

***Proof*** We can apply the general comparison principle of Proposition 2.68. In Example 2.67 we have shown that Assumption 2.65 holds if $\psi_1 \geq \psi_2$ and $g_1 \geq g_2$. Assumption 2.65 holds as can be seen by a simple modification of Example 2.66. Hence, Proposition 2.68 leads to the stated result. □

Proposition 2.71 answers questions regarding the change of the value of the option if, for example, the pay-off (corresponding to the obstacle) or the intermediate payments (corresponding to $f$) are modified. For example, Proposition 2.71 shows that the increase of the pay-off or payments will lead to an increase in the value of the option, as expected. Similar results may be obtained by changing parameters in the operator $A$, e.g., the interest rate $r$.

There is more than the abstract theory can offer us:

**Proposition 2.72** *In the case of a single asset* ($N = 1$)*, the function* $u$ *is a continuous function.*

***Proof*** Now $D = (-L, L) \subset \mathbb{R}$, an interval. We already know that $u \in W^{1,2}(D)$ by the existence result. By the Sobolev embedding (Theorem 2.48(i)), $W^{1,2}(D) \subset C(D)$ (with continuous injection); hence $u$ is continuous. □

We have obtained important information so far, but there are still interesting questions that are left unattended, such as:

(a) Is the function $u$ continuous for basket options, i.e., in the case $N \geq 2$?
(b) Can we guarantee that $u$ is differentiable in a stronger sense than the one implied by the fact that $u \in W^{1,2}(D)$?
(c) Does the solution $u$ of the VI (2.63) (even for $N = 1$) satisfy the linear complementarity problem

$$Au - f \geq 0, \quad u \geq \psi, \quad (Au - f)(u - \psi) = 0 \tag{2.65}$$

and the free boundary formulation, which is where we actually started from and that allows us to find a stopping criterion?

The answer to these questions requires more than what we already have: in particular information about higher weak derivatives than the first. This leads us to the intriguing question of regularity of weak solutions.

### 2.6.6 Regularity

We now turn to the question of regularity which is very important in practical applications of VIs in finance and management science. The question of regularity is related to whether the solution of the VI (2.45), obtained in Sobolev spaces, can actually get any better, i.e., continuous or smoother under extra assumptions on the data of the problem. Such questions can be tackled using the Sobolev embeddings. In particular if one can show membership of the solution in higher order Sobolev spaces, then the Sobolev embedding theorem 2.48 allows us to deduce that the solution is continuous or even continuously differentiable (see also Remark 2.54).

The membership of the solution in higher Sobolev spaces typically follows using a priori bounds for $u$ or its derivatives. For example, if we can show that $Au \in L^2(D)$ (rather than just $Au \in (W_0^{1,2}(D))'$, as is guaranteed by the theory so far), then the second order weak derivatives of $u$ are in $L^2(D)$ and hence will guarantee that $u \in W^{2,2}(D)$. This will allow to apply the arguments of Remark 2.54 to obtain regularity.

There are various approaches to the problem or regularity. One of the first approaches was via the method of penalization (see, e.g., Bensoussan and Lions (2011)), which was introduced also as a method for proving existence of solutions and can further be extended to prove regularity. Other methods have also been proposed (see, e.g., Rodrigues (1987)), based on duality arguments, which require at points some rather subtle functional analytic arguments. Here we favour the first approach.

The basic concept of penalization is to substitute the inequality with a differential equation with a penalty term which becomes increasingly large if the solution of the differential equation is not in $K$. The penalty term essentially forces the solution of the equation to lie in $K$.

**Do Remember 2.73 (Penalization Method)** *Replace the VI*

$$u \in K : \quad \langle Au - f, v - u \rangle \geq 0, \ \forall \, v \in K, \tag{2.66}$$

*with the family of equations* $Au_\epsilon + \beta_\epsilon(u_\epsilon) = f$, $\epsilon > 0$, *whose weak form is*

$$u_\epsilon \in X \ : \ \underbrace{\langle Au_\epsilon, v \rangle}_{=\mathbf{a}(u_\epsilon, v)} + \langle \beta_\epsilon(u_\epsilon), v \rangle - \langle f, v \rangle = 0, \quad \forall \, v \in X, \tag{2.67}$$

*for a sequence of (monotone) penalty functions* $(\beta_\epsilon = \frac{1}{\epsilon}\beta \ : \ \epsilon > 0)$, *with* $\beta : X \to X$, *Lipschitz continuous, such that*

$$\begin{aligned} &\beta(v) = 0 \iff v \in K, \quad (Ker(\beta) = K), \\ &\langle \beta(v_1) - \beta(v_2), v_1 - v_2 \rangle \geq 0. \end{aligned} \tag{2.68}$$

**Example 2.74** A good candidate for a penalty function can be $\beta(v) = (I - \Pi_K)(v)$. For example, when $K = \{v \in W_0^{1,2}(D) \ : \ v \geq \psi\}$, in which case $\Pi_K = \psi + (v - \psi)^+$, a suitable penalty function is provided by $\beta(v) = v - \psi - (v - \psi)^+ = -(v - \psi)^-$. Properties (2.68) can be checked directly. ◁

The method can work in the context of general Hilbert spaces $X$, in particular in the context of Gelfand triples $X \subset X_0 \subset X'$, and even in the general Banach space setting for linear or non-linear operators.

Here, for simplicity, we focus on the case where $A$ is a linear operator in the form of Definition 2.55, and $X = W_0^{1,2}(D)$, $X_0 = L^2(D)$, with $K = \{v \in W_0^{1,2}(D) \ : \ v \geq \psi\}$. Its extension to other settings is left to the reader.

**Remark 2.75** The key idea behind the penalty method is the simple observation that for any $v \in K$ the penalized equation can be written as

$$Au_\epsilon + \beta_\epsilon(u_\epsilon) - \beta_\epsilon(v) - f = 0, \tag{2.69}$$

so upon taking the weak form with a test function $u_\epsilon - v$ the penalty term keeps a sign on account of monotonicity, hence providing useful estimates for the term $Au_\epsilon$. These estimates can be used to show that as $\epsilon \to 0$, the effect of the penalty term is to force $u_\epsilon \to u \in K$. ⊕

By the properties of the penalty function, it can be shown that the solutions of (2.67) converge to the solutions of (2.66) in the limit as $\epsilon \to 0$, i.e., we may approximate the solutions of the VI (2.66) by solutions of equation (2.67) for sufficiently small $\epsilon$. Since equations are easier to solve (both analytically and numerically) than inequalities, this is an important step in its own right. But, having an approximating equation can have other benefits as well. For example, for any solution $u_\epsilon$ of (2.67), the equation yields that $Au_\epsilon = f - \beta_\epsilon(u_\epsilon)$, so the monotonicity of $\beta_\epsilon$ can allow us to obtain a priori bounds for $Au_\epsilon$ which will lead to membership of $u_\epsilon$ in higher order Sobolev spaces. Then passing to the limit in $\epsilon$, we recover the necessary bounds for $u$ and hence the required regularity.

The following proposition shows that the approximation via penalization is indeed successful.

**Proposition 2.76** *Assume that* $\mathbf{a} : X \times X \to \mathbb{R}$, $X = W_0^{1,2}(D)$, *is continuous and coercive. Let* $u_\epsilon$ *be the solution of the penalized equation* (2.67) *for any* $\epsilon > 0$ *and* $u$ *the solution of the VI* (2.66). *Then (up to subsequences) the sequence* $(u_\epsilon)$ *has a strong limit in* $W_0^{1,2}(D)$ *which is the solution of* (2.66), *i.e.,*

$$\|u_\epsilon - u\|_{W_0^{1,2}(D)} \to 0, \quad as \ \epsilon \to 0. \tag{2.70}$$

***Proof (Sketch)*** The proof is essentially based on Remark 2.75.

0. An essential step in the proof is an a priori bound for the solution (given that it exists). If $u_\epsilon \in X$ is the weak solution of (2.67), then for any $v \in K$ we can set as test function in (2.67) $v - u_\epsilon \in X$, and this (after subtracting $\beta_\epsilon(v) = 0$ from the equation) yields

$$\begin{aligned}
&\mathbf{a}(u_\epsilon, v - u_\epsilon) - \langle f, v - u_\epsilon \rangle = \underbrace{\langle \beta_\epsilon(v) - \beta_\epsilon(u_\epsilon), v - u_\epsilon \rangle}_{\geq 0, \text{ by } (2.68)} \geq 0 \implies \\
&\mathbf{a}(u_\epsilon, u_\epsilon) \leq \mathbf{a}(u_\epsilon, v) - \langle f, v - u_\epsilon \rangle, \quad \epsilon > 0.
\end{aligned} \tag{2.71}$$

We can then estimate the LHS using coercivity, and the RHS using continuity of the form $\mathbf{a}$, to obtain

$$\begin{aligned}
c_0 \|u_\epsilon\|^2_{W_0^{1,2}(D)} &\leq c_1 \|v\|_{W_0^{1,2}(D)} \|u_\epsilon\|_{W_0^{1,2}(D)} + \|f\|_{L^2(D)} \|v - u_\epsilon\|_{L^2(D)} \\
&\leq c_1 \|v\|_{W_0^{1,2}(D)} \|u_\epsilon\|_{W_0^{1,2}(D)} + c_P \|f\|_{L^2(D)} \|v - u_\epsilon\|_{W_0^{1,2}(D)},
\end{aligned}$$

where for the last step we have used the Poincaré inequality. Using the triangle inequality for $\|v - u_\epsilon\|_{W_0^{1,2}(D)}$ and the fact that $\|f\|_{L^2(D)}$ and $\|v\|_{W_0^{1,2}(D)}$ are bounded independently of $\epsilon$, we conclude that

$$c_0 \|u_\epsilon\|^2_{W_0^{1,2}(D)} \leq c_2 \|u_\epsilon\|_{W_0^{1,2}(D)} + c_3, \quad \epsilon > 0, \tag{2.72}$$

with $c_2, c_3$ constants depending on $\|f\|_{L^2(D)}$ and $\|v\|_{W_0^{1,2}(D)}$, importantly independent of $\epsilon$. Then (2.72) implies that

$$\|u_\epsilon\|_{W_0^{1,2}(D)} \leq c, \quad \epsilon > 0, \quad c \text{ independent of } \epsilon. \tag{2.73}$$

These bounds are a priori bounds, i.e., properties of the solution given that a solution exists. We will leave the question of existence for the moment and proceed to further properties of $u_\epsilon$. We only state here that the actual proof of existence is (a) either based upon a finite dimensional approximation, the a priori limits and a weak compactness argument (see, e.g., Bensoussan and Lions (2011)) or (b) upon a fixed point approach using the Lions-Stampacchia abstract existence proof.
1. To show the existence of a limit in $K$, using the a priori bounds of step 0, leading to (2.73), by a weak compactness argument (see Do Remember 2.49), there exists $u \in W_0^{1,2}(D)$ such that (up to subsequences) $u_\epsilon \rightharpoonup u$ in $W_0^{1,2}(D)$. Moreover, by the compact embedding (up to a further subsequence), $u_\epsilon \to u$ in $L^2(D)$.

To show that $u \in K$, we start again from (2.71), move the term $\langle \beta_\epsilon(u_\epsilon), u_\epsilon - v \rangle$ to the LHS, and use (2.73), which leads us to the boundedness of all the remaining terms, uniformly in $\epsilon$, to obtain

$$0 \leq \langle \beta(u_\epsilon) - \beta(v), u_\epsilon - v \rangle \leq c\,\epsilon,$$

for any $v \in K$, with $c$ independent of $\epsilon$. Passing to the limit[18] as $\epsilon \to 0$, we obtain that $\beta(u) = 0$; hence $u \in K$.
2. To establish that the convergence $u_\epsilon \to u$ is strong in $W_0^{1,2}(D)$, we require coercivity once more:

$$
\begin{aligned}
c_0 \|u - u_\epsilon\|^2_{W_0^{1,2}(D)} &\leq \mathbf{a}(u - u_\epsilon, u - u_\epsilon) = \mathbf{a}(u, u - u_\epsilon) - \mathbf{a}(u_\epsilon, u - u_\epsilon) \\
&\overset{(2.71)}{=} \mathbf{a}(u, u - u_\epsilon) - \langle f, u - u_\epsilon \rangle - \underbrace{\langle \beta_\epsilon(u_\epsilon) - \beta_\epsilon(u), u_\epsilon - u \rangle}_{\geq 0} \\
&\leq \mathbf{a}(u, u - u_\epsilon) - \langle f, u - u_\epsilon \rangle
\end{aligned}
$$

(where we used (2.71) for $v = u \in K$). Since the RHS of the above tends to 0 (by $u_\epsilon \rightharpoonup u$ in $W_0^{1,2}(D)$) as $\epsilon \to 0^+$, strong convergence in $W_0^{1,2}(D)$ is established.
3. We now show that the limit $u \in K$ solves (2.66). To see this take any $v \in K$, and use as test function in the weak formulation of (2.67) (equiv. (2.69)) $v - u_\epsilon$. Since $\beta_\epsilon(v) = 0$ because $v \in K$, rearranging the weak form by simply inserting this null term yields

$$
\mathbf{a}(u_\epsilon, v - u_\epsilon) - \langle f, v - u_\epsilon \rangle = \underbrace{\langle \beta_\epsilon(v) - \beta_\epsilon(u_\epsilon), v - u_\epsilon \rangle}_{\geq 0, \text{ by } (2.68)} \geq 0. \tag{2.74}
$$

Letting $\epsilon \to 0^+$ (along the convergent subsequence) and using step 2 (since the convergence of $\mathbf{a}(u_\epsilon, u_\epsilon)$ requires strong convergence), we obtain the VI. □

We can now obtain our first regularity result. To fix ideas let us restrict to $K$ as in Example 2.74.

**Proposition 2.77** *Assume that the coefficients of the operator $A$ are sufficiently smooth,*[19] *$f \in L^2(D)$, and $K$ as in Example 2.74 with $A\psi \in L^2(D)$. Then, the solution of the VI* (2.66) *satisfies $u \in W^{2,2}(D)$ and satisfies the bound*

$$
\|Au\|_{L^2(D)} \leq c(\|f\|_{L^2(D)} + \|A\psi\|_{L^2(D)}),
$$

*for some constant $c$.*

***Proof (Sketch)*** The idea of the proof is simple: By the properties of the penalty term $\beta_\epsilon(u_\epsilon) \in L^2(D)$. Since by the penalized equation $Au_\epsilon = f - \beta_\epsilon(u_\epsilon)$ and $f \in L^2(D)$, we see that $Au_\epsilon \in L^2(D)$. By the form of the operator $A$, this shows that $u_\epsilon \in W^{2,2}(D)$. It remains to show that passing to the limit as $\epsilon \to 0$, we get

---

[18] This step requires some caution as the convergence $u_\epsilon \rightharpoonup u$ is still weak. One needs to take the lim inf in the above and use the particular form of $\beta$ (see Example 2.74) and the weak lower semicontinuity property (see Do Remember 2.49).

[19] See step 0 of proof for details.

that $Au \in L^2(D)$, and hence $u \in W^{2,2}(D)$. These two steps can be made rigorous as follows:
0. We recall the regularity theory for linear elliptic problems of the form $Au = w$, according to which roughly speaking the solution $u$ gains two derivatives, as compared to $w$. More precisely (see, e.g., Theorem 2, Ch. 6. Sec. 3 p. 332 in Evans (2010)) if $a_{ij}, b_i, c \in C^1(D)$ and $w \in L^2(D)$, then the weak solution of the elliptic PDE, $Au = w$ in $D$, admits a solution $u \in W^{2,2}_{loc}(D)$ with the estimate

$$\|u\|_{W^{2,2}(D_0)} \leq c(\|w\|_{L^2(D)} + \|u\|_{L^2(D)}), \tag{2.75}$$

for every $D_0 \subset\subset D$, with the constant $c$ depending only on $D, D_0$ and the coefficients of $L$. If $a_{ij}, b_i, c \in C^1(\bar{D})$ and $\partial D \in C^2$, then for the homogeneous Dirichlet boundary condition the solution enjoys the stated smoothness up to the boundary, and estimate (2.75) can be extended for $D_0 = D$ (see Theorem 5, op. cit.).
1. Let $w = f - \beta_\epsilon(u_\epsilon) \in L^2(D)$. We will then use the regularity theory of the weak form of the equation $Au_\epsilon = w$ to obtain that $u_\epsilon \in W^{2,2}(D)$. This step may require the use of a fixed point argument or a finite dimensional approximation scheme. Both this step and passing to the limit as $\epsilon \to 0^+$ require some appropriate a priori bounds for the solution.
2. Since $f$ is given, some control is required on the norm $\|\beta_\epsilon(u_\epsilon)\|_{L^2(D)}$. This can be done using $v = \beta_\epsilon(u_\epsilon)$ in the weak formulation of the penalized problem, which will generate the norm $\|\beta_\epsilon(u_\epsilon)\|^2_{L^2(D)}$, a step that can be justified as long as $\beta_\epsilon(u_\epsilon) \in W^{1,2}_0(D)$. Under this assumption,[20] (2.67) yields

$$\mathbf{a}(u_\epsilon, \beta_\epsilon(u_\epsilon)) + \|\beta_\epsilon(u_\epsilon)\|^2_{L^2(D)} = \langle f, \beta_\epsilon(u_\epsilon)\rangle. \tag{2.76}$$

Clearly we need to control the first term. This requires some knowledge of the nature of the set $K$. For $K$ as in Example 2.74,

$$\begin{aligned} \mathbf{a}(u_\epsilon, \beta_\epsilon(u_\epsilon)) &= \frac{1}{\epsilon}\mathbf{a}(u_\epsilon - \psi, -(u_\epsilon - \psi)^-) + \frac{1}{\epsilon}\mathbf{a}(\psi, -(u_\epsilon - \psi)^-) \\ = -\frac{1}{\epsilon}\underbrace{\mathbf{a}((u_\epsilon - \psi)^+, (u_\epsilon - \psi)^-)}_{=0} &+ \frac{1}{\epsilon}\underbrace{\mathbf{a}((u_\epsilon - \psi)^-, (u_\epsilon - \psi)^-)}_{\geq 0} - \frac{1}{\epsilon}\mathbf{a}(\psi, (u_\epsilon - \psi)^-) \\ &\geq \mathbf{a}(\psi, \beta_\epsilon(u_\epsilon)) = \langle A\psi, \beta_\epsilon(u_\epsilon)\rangle = \langle A\psi, \beta_\epsilon(u_\epsilon)\rangle_{L^2(D)}, \end{aligned} \tag{2.77}$$

where we used (in this order) bilinearity, the fact that $\mathbf{a}(v^+, v^-) = 0$ for any $v \in W^{1,2}_0(D)$, coercivity, and finally the definition of operator $A$ in terms of the bilinear form. The final equality comes from the assumption that $A\psi \in$

[20] See Proposition 2.76 and the form of the penalty term.

$L^2(D)$, which allows us to interpret the duality pairing $\langle \cdot, \cdot \rangle$ between the spaces $W_0^{1,2}(D)'$, $W_0^{1,2}(D)$ as the inner product of $L^2(D)$ by recalling that $W_0^{1,2}(D) \subset L^2(D) \subset W_0^{1,2}(D)'$.

Using the above in (2.76), we obtain that

$$\begin{aligned} \|\beta_\epsilon(u_\epsilon)\|^2_{L^2(D)} &= \langle f, \beta_\epsilon \rangle - \mathbf{a}(u_\epsilon, \beta_\epsilon(u_\epsilon)) \overset{(2.77)}{\leq} \langle f, \beta_\epsilon \rangle - \langle A\psi, \beta_\epsilon(u_\epsilon) \rangle \\ &= \langle f - A\psi, \beta_\epsilon(u_\epsilon) \rangle \leq \|f - A\psi\|_{L^2(D)} \|\beta_\epsilon(u_\epsilon)\|_{L^2(D)}, \end{aligned} \tag{2.78}$$

and the desired estimate

$$\|\beta_\epsilon(u_\epsilon)\|_{L^2(D)} \leq \|f - A\psi\|_{L^2(D)}$$

follows upon dividing both sides by $\|\beta_\epsilon(u_\epsilon)\|_{L^2(D)}$. Note that the bound is independent of $\epsilon$. Hence, the penalized equation yields that $Au_\epsilon \in L^2(D)$, with an (a priori) bound of the form

$$\|Au_\epsilon\|_{L^2(D)} \leq c(\|f\|_{L^2(D)} + \|A\psi\|_{L^2(D)}),$$

with $c$ a finite constant independent of $\epsilon$, and from that a bound $\|u_\epsilon\|_{W^{2,2}(D)} \leq c$ with $c$ independent of $\epsilon$.

3. Obtaining that $u \in W^{2,2}(D)$ requires a careful passage to the limit. Since $u_\epsilon$ is uniformly bounded (in $\epsilon$), there exists a $\bar{u} \in W^{2,2}(D)$ such that (up to subsequences) $u_\epsilon \rightharpoonup \bar{u}$ in $W^{2,2}(D)$. On the other hand, we know from Proposition 2.76 that there exists a $u \in W^{1,2}(D)$ such that (up to subsequences) $u_\epsilon \rightharpoonup u$ in $W^{1,2}(D)$ which is also the solution of the VI. Passing to a common subsequence and employing a uniqueness of limits argument (by first taking both limits in the sense of distributions), we conclude that $u = \bar{u} \in W^{2,2}(D)$. Passing to the limit, by the weak lower semicontinuity of the norm (see Do Remember 2.49), we obtain the result. □

**Remark 2.78** Note that estimate (2.78) can be sharpened since

$$\begin{aligned} &\|\beta_\epsilon(u_\epsilon)\|^2_{L^2(D)} \leq \langle f - \langle A\psi, \beta_\epsilon(u_\epsilon) \rangle = \langle -(A\psi - f)^+ + (A\psi - f)^-, \beta_\epsilon(u_\epsilon) \rangle \\ &= -\langle (A\psi - f)^+, \beta_\epsilon(u_\epsilon) \rangle + \underbrace{\langle (A\psi - f)^-, \beta_\epsilon(u_\epsilon) \rangle}_{\leq 0, \text{ since } \beta_\epsilon(u_\epsilon) \leq 0} \leq -\langle (A\psi - f)^+, \beta_\epsilon(u_\epsilon) \rangle, \end{aligned}$$

leading to the bound

$$\|Au\|_{L^2(D)} \leq \|f\|_{L^2(D)} + \|(A\psi - f)^+\|_{L^2(D)},$$

resembling the Lewy-Stampacchia estimates, who established an estimate of this type in Lewy and Stampacchia (1970) for the Laplace operator. ⊕

**Remark 2.79 (Higher Regularity)** One may obtain higher regularity using the above scheme by assuming higher regularity of the data, building on regularity results of linear problems which extend (2.75) for $\|u\|_{W^{2,p}(D)}$, $p \geq 2$, as long as $w \in L^p(D)$. By using $v = |\beta_\epsilon(u_\epsilon)|^{p-2}\beta_\epsilon(u_\epsilon)$ as test function in (2.67) (the weak formulation of the penalized problem), we may obtain a priori estimates of the form (2.78) for $\|\beta_\epsilon(u_\epsilon)\|_{L^p(D)}$ in terms of $\|f\|_{L^p(D)}$, $\|A\psi\|_{L^p(D)}$.[21] Then, using the regularity results for elliptic systems, we may use this bound to obtain the solvability of the penalized problem, and by passing to the limit as $\epsilon \to 0$, using essentially the same arguments as above, in $W^{2,p}(D)$, $p \in [2, \infty)$, as long as the data satisfy $f, A\psi \in L^p(D)$. The details are left to the reader, see also Chipot (1984). Note the interesting pattern that appears: The solution is as smooth as the obstacle. For more on regularity, see, e.g., Fernández-Real and Ros-Oton (2023). ⊕

### 2.6.7 *Fully Worked Example: Perpetual American Basket Options Revisited*

Having obtained the regularity results we may now revisit the example of perpetual American basket options with respect to continuity issues or return to the question of whether the VI admits a linear complementarity formulation. The results and techniques of Sect. 2.6.6 can be directly applied to the problem of the valuation of perpetual American basket options, for suitably smooth pay-off functions. For simplicity we may consider the case of a barrier basket option, designed such that $u(x) = 0$ on $\partial D$, the valuation problem for which reduces to (2.63) for $g = 0$.

**Proposition 2.80** *Assume that the matrix $\sigma$ is positive definite and that the final pay-off (obstacle) $\psi$ satisfies $A\psi \in L^p(D)$ and the intertemporal pay-off $f \in L^p(D)$ for $p \in [2, \infty)$. Then, the value function $u \in W^{2,2}(D)$ satisfying the bound*

$$\|Au\|_{L^p(D)} \leq \|f - A\psi\|_{L^p(D)} + \|f\|_{L^p(D)},$$

*If $N < 2p$, the solution of the VI is continuous. If $p > N$, then $u \in C^1(\bar{D})$.*

***Proof*** Since $\sigma$ is positive definite, the corresponding bilinear form **a** is continuous and coercive. The regularity and the corresponding bound then follows as a simple application of Proposition 2.76 and Remark 2.79.

The regularity guaranteed by the first part of the proposition allows us to obtain the continuity of $u$. Reasoning as in Remark 2.54, we see that if $p < N$ but $p > N/2$, then the fact that $u \in W^{2,p}(D)$ implies that $Du \in W^{1,Np/(N-p)} \subset C(\bar{D})$. In our case we have that $p = 2$, so this reasoning can be justified as long as $N = 2$; that is, if the number of assets is up to 2, then the solution of the VI that yields the

[21] A technical truncation approximation argument will be needed here, see, e.g., Chipot (1984).

option price is continuous. With a similar argumentation, if $p > N$, then $Du \in W^{1,p}(D) \subset C(\bar{D})$; hence $u \in C^1(\bar{D})$. □

Moreover, the regularity results allow us to show that $u$ solves the linear complementarity formulation (2.65) or its equivalent free boundary formulations.

$$\min(Au - f, u - \psi) = 0 \iff Au - f \geq 0,\ u - \psi \geq 0,\ (Au - f)(u - \psi) = 0$$

$$\iff \begin{cases} Au = f & \text{if } u > \psi, \\ Au > f & \text{if } u = \psi, \\ u \geq \psi, & \text{a.e. in } D, \end{cases} \iff \begin{cases} Au = f & \text{on } \mathscr{C}^c, \\ Au > f & \text{on } \mathscr{C} := \{x \in D \,:\, u(x) = \psi(x)\}, \\ u \geq \psi, & \text{a.e. in } D, \end{cases} \tag{2.79}$$

with all equalities and inequalities interpreted a.e. in $D$ and supplemented with $u = 0$ on $\partial D$ (interpreted in the trace sense). Note that the coincidence set $\mathscr{C} := \{x \in D \,:\, u(x) = \psi(x)\}$ is unknown and only determined when the solution is known. The set $\mathscr{C}$ is called the free boundary to distinguish from the boundary $\partial D$ where the value of the solution is known too, but its location is known a priori and not determined by the solution itself.

**Proposition 2.81** *Under the assumptions of Proposition 2.80, the formulations* (2.45) *and* (2.79) *are equivalent.*

***Proof*** Follows from Proposition 1.25, noting that the integrals involved are well-posed as long as $Au \in L^2(D)$. □

## 2.7 Parabolic Variational Inequality Problems

We now turn to generalize the results of Sect. 2.6 to the case of time-dependent VIs of the type

$$\begin{gathered} \frac{\partial u}{\partial t} + Au - f \geq 0,\ \ u - \psi \geq 0,\ \ (\frac{\partial u}{\partial t} + Au - f)(u - \psi) = 0, \\ u(t, x) = 0,\ \text{ on } \partial D,\ \ u(0, x) = u_0(x) \text{ on } D, \end{gathered} \tag{2.80}$$

where $A$ is an operator of the general class introduced in Definition 2.55. The various concepts and definitions introduced in Sect. 2.6 will be needed for the time-dependent (parabolic) case as well. The major difference here is that $u : [0, T] \times D \to \mathbb{R}$ is now a function of time as well, and the problem is also supplemented with an initial condition. A good way to conceptualize the problem is to "separate" the temporal from the spatial part and consider for each $t \in [0, T]$ a function $u(t, \cdot) : D \to \mathbb{R}$, which carries all the spatial information concerning the solution of the problem at time $t$; this is considered as an element of an appropriate function space $X$, containing the effect of the boundary conditions. A variety of possible boundary conditions can be accommodated in this way.[22] With the above conceptualization the solution $u : [0, T] \times D \to \mathbb{R}$ to (2.80) can be understood as a function $u : [0, T] \to X$, a function from $[0, T]$ to a suitable function space $X$ (which carries the spatial information of the solution—hence the boundary conditions) such that $u(0) = u_0 \in X$, so that the initial condition is also satisfied. In some sense using this idea we may turn a PDE into an ordinary differential equation in a properly selected function space.

With this introduction at hand it makes sense to try for a weak formulation of (2.80). Naively thinking, we can consider (2.80) as the elliptic case with an extra term coming in, $\frac{\partial u}{\partial t}$, that of course is somehow connected (through the derivative operator) with the function $u$ showing in the rest of the problem. We can then provide a weak formulation in terms of

$$u \in \mathscr{K} \ : \ \langle \frac{\partial u}{\partial t} + Au - f, v - u \rangle \geq 0, \ a.e. \ [0, T]. \ \forall \ v \in \mathscr{K}, \tag{2.81}$$

for a suitably selected convex set $\mathscr{K}$ carrying the information $u(t, x) \geq \psi$ in some appropriate sense. This is a good plan, but there are some loose ends we must fix before embarking into it. That involves questions, such as what type of mathematical object $\frac{\partial u}{\partial t}$ is, how can we select the set $\mathscr{K}$, etc., all of which return to the same starting point: How do we understand a function $u : [0, T] \to X$, where $X$ is a function space. Moreover, the naive formulation (2.81) entails delicate points such as, for instance, that $u$ and $\frac{\partial u}{\partial t}$ may not belong to the same function space.

Hence, before continuing with (2.81), we need to introduce some mathematical preliminaries first.

### 2.7.1 *Interlude: Spaces of Vector-Valued Functions*

In this section we collect some fundamental facts on vector-valued functions of the form $u : [0, T] \to X$, where $X$ is a function space, equipped with a norm that makes $X$ a Hilbert space or a Banach space.

[22] For example, $X = W_0^{1,2}(D)$ for homogeneous Dirichlet boundary conditions, $X = W_0^{1,2}(D) + \bar{g}$ with a suitable $\bar{g} = \mathscr{L}(g)$ for non-homogeneous Dirichlet boundary conditions $u = g$ on $\partial D$, or $X = W^{1,2}(D)$ for Neumann boundary conditions.

We need the following definitions:

**Definition 2.82** Define $L^2((0,T);X) := \{u : [0,T] \to X : \int_0^T \|u(t)\|_X^2 dt < \infty\}$. This is a Hilbert space equipped with the norm

$$\|u\|_{L^2((0,T);X)} = \left(\int_0^T \|u(t)\|_X^2 dt\right)^{1/2}.$$

**Definition 2.83** Let $X, Y$ be two Hilbert spaces such that $X \subset Y$ (continuous embedding), and consider $u \in L^2((0,T);X)$. We define the derivative $u' \in L^2((0,T);Y)$ as $u' = v$ with $v \in L^2((0,T);Y)$ such that

$$-\int_0^T u(t)\phi'(t)dt = \int_0^T v(t)\phi(t)dt, \quad \forall\ \phi \in C_c^1([0,T]).$$

The above definition makes sense as an equality in $Y$, with the LHS and the RHS being consistent since $X \subset Y$.

A particularly important case is the case where $Y = X'$, the dual space of $X$.

**Example 2.84** The above definitions are not mathematical artifacts but necessities. Consider $u(t,x) = \sum_n (a_n e^{-n^2 t} + n^{-2} f_n)\sin(nx)$, the solution of $u_t - u_{xx} = f$ for $f(x) = \sum_n f_n \sin(nx)$, and initial condition $u_0$ given by $u_0 = \sum_n a_n \sin(nx)$ in the domain $D = [-\pi, \pi]$. If $(f_n)_{n\in\mathbb{N}}$ satisfies $\sum_n |f_n|^2 < \infty$, then $f \in L^2(D)$, and if $(a_n)_{n\in\mathbb{N}}$ satisfies $\sum_n n^2|a_n|^2 < \infty$, then $u_0 \in W_0^{1,2}(D)$. We can see that $u(t) \in W_0^{1,2}(D)$. However $u_t(t,x) = -\sum_n e^{-n^2 t} n^2 a_n \sin(nx)$, for which $u_t \notin W_0^{1,2}(D)$ but rather $u_t \in (W_0^{1,2}(D))'$. ◁

The setting $u \in L^2(0,T;X)$ and $u' \in L^2(0,T;X')$ in Definition 2.83 can be used within the general framework of Gelfand triples (see Sect. 2.6.2), which was very useful in the study of the elliptic case. From now on, we will assume the embeddings $X \subset X_0 \subset X'$ (in our standard examples $X = W_0^{1,2}(D)$, $X_0 = L^2(D)$) and consider functions $u \in L^2(0,T;X)$ with derivatives $u' = \frac{\partial u}{\partial t} \in L^2(0,T;X')$, in the sense of Definition 2.83. In this setting an important result is that

$$\begin{aligned} &\int_0^T \langle \frac{\partial w}{\partial t}(t,\cdot), v(t,\cdot)\rangle dt + \int_0^T \langle \frac{\partial v}{\partial t}(t,\cdot), w(t,\cdot)\rangle dt \\ &= \langle w(T,\cdot), v(T,\cdot)\rangle - \langle w(0,\cdot), v(0,\cdot)\rangle, \end{aligned} \tag{2.82}$$

as long as all the quantities in (2.82) are well defined. Note that the integral on the LHS is well defined if $\frac{\partial w}{\partial t}, \frac{\partial v}{\partial t} \in L^2(0,T;(W_0^{1,2}(D))')$ and $w, v \in L^2(0,T;W_0^{1,2}(D))$ or better. In the special case where $w = v$, the RHS can be interpreted in terms of the norms $\|u\|^2_{L^2(D)}, \|v\|^2_{L^2(D)}$. This result essentially follows by the properties of multiple integrals. Moreover, it is possible to show that

$$
\begin{aligned}
&u \in L^2(0,T;W_0^{1,2}(D)), \ \frac{\partial u}{\partial t} \in L^2(0,T;(W_0^{1,2}(D))') \\
&\quad \Longrightarrow u \in C([0,T];L^2(D)), \\
&\max_{0\le t\le T} \|u(t)\|_{L^2(D)} \le c\left( \|u\|_{L^2(0,T;W_0^{1,2}(D))} + \left\| \frac{\partial u}{\partial t} \right\|_{L^2(0,T;(W_0^{1,2}(D))')} \right), \qquad (2.83)
\end{aligned}
$$

with a constant $c$ depending only on $T$ (see Theorem 3, Section 5.9 Evans (2010)). This is a very useful result as it allows us to consider functions $u \in L^2(0,T;W_0^{1,2}(D)), \frac{\partial u}{\partial t} \in L^2(0,T;(W_0^{1,2}(D))')$ as continuous functions on $L^2(D)$. In general we have the celebrated Aubin-Lions lemma (see, e.g., Aubin (1963) or Roubicek (2013)):

**Theorem 2.85 (Aubin-Lions Lemma)** *Let $X_1, X, X_{-1}$ be Banach spaces such that $X_1 \overset{c}{\hookrightarrow} X \hookrightarrow X_{-1}$ (with $\overset{c}{\hookrightarrow}$ meaning continuous and compact embedding). Consider the space*

$$
Z := \left\{ w \in L^p(0,T;X_1) \ : \ \frac{\partial w}{\partial t} \in L^q(0,T;X_{-1}) \right\},
$$

*where $p, q \in [1,\infty]$.*

*(i) If $p < \infty$, then the embedding $Z \overset{c}{\hookrightarrow} L^p(0,T;X)$ holds.*
*(ii) If $p = \infty$ and $q > 1$, then the embedding $Z \overset{c}{\hookrightarrow} C(0,T;X)$ holds.*

### 2.7.2 The Formulation of a Class of Parabolic VIs

We are now ready to introduce the weak formulation for parabolic VIs in Hilbert spaces.

**Definition 2.86** Let $X$ be a Hilbert space, $K \subset X$ be a convex, closed set, and $\mathbf{a} : X \times X \to \mathbb{R}$ be a bilinear form with an associated operator $A : X \to X'$ of the type introduced in Definition 2.6 and the corresponding VI

$$
\begin{aligned}
u \in \mathcal{K} \ : \ & \int_0^T \left( \left\langle \frac{\partial u}{\partial t}, v - u \right\rangle + \mathbf{a}(u, v - u) \right) dt \\
& \geq \int_0^T \langle f, v - u \rangle dt, \ \ \forall \, v \in \mathcal{K}, \\
& \mathcal{K} := \{ u \in L^2(0, T; X), \ \ s.t. \ u(t, \cdot) \in K, \ \ a.e. \ t \in (0, T) \}, \\
& u(0) = u_0,
\end{aligned}
\tag{2.84}
$$

called the weak form of the parabolic VI (2.80).

The time dependence induces the need to introduce integration not only over $x \in D$ but also over $t \in [0, T]$ when introducing the weak formulation. Essentially the above means that the weak formulation (2.81) holds a.e. in $[0, T]$. Apart from that, the motivation and intuition remains the same. As in the case of the elliptic problem, here we will consider either $X = W^{1,2}(D)$ or $X = W_0^{1,2}(D)$.

### 2.7.3 *Existence and Regularity: Penalization*

To study the existence and regularity for parabolic VIs penalization proves to be a suitable method. An alternative would be to try to apply the abstract results in the spirit of the Lions-Stampacchia approach using essentially monotonicity arguments. This is feasible but considerably more technical than the approach via penalization and would require rather advanced techniques from the theory of monotone operators.

**Do Remember 2.87 (Penalization for Parabolic VIs)** *The idea of penalization for parabolic problems is similar as for elliptic problems: We replace the VI* (2.84) *by a penalized version, in terms of the family of equations*

$$
\text{Find } u_\epsilon \in L^2(0, T; X), \ \frac{\partial u_\epsilon}{\partial t} \in L^2(0, T; X'), \ u_\epsilon(0) = u_0 :
$$

$$
\int_0^T \left( \left\langle \frac{\partial u_\epsilon}{\partial t}, v \right\rangle + \mathbf{a}(u_\epsilon, v) + \langle \beta_\epsilon(u_\epsilon), v \rangle \right) dt = \int_0^T \langle f, v \rangle dt, \ \forall \, v \in L^2(0, T; X),
\tag{2.85}
$$

(continued)

**Do Remember 2.87** (continued)
*where $(\beta_\epsilon)_{\epsilon>0}$ is an appropriate family of penalty functions as in (2.68). We then appropriately pass to the limit as $\epsilon \to 0$ to show that this limit is a solution of the VI (2.84). The regularity of the solution of (2.84) follows by the regularity of the solution $u_\epsilon$ of (2.85) again by appropriate passage to the limit.*

Note that the well-developed regularity theory for parabolic equations can be of great help in this task. The method is general, but here for concreteness we will focus on the same case as we did for the elliptic case in Sect. 2.6.6.

**Existence**

**Proposition 2.88** *Under the extra assumption that $\frac{\partial \psi}{\partial t}, A\psi \in L^2(0,T;L^2(D))$, the penalized problem (2.85) admits a unique solution $u_\epsilon$. The sequence $(u_\epsilon)$ (up to subsequences) has a strong limit in $L^2(0,T;W_0^{1,2}(D))$ which is the solution of (2.84), i.e.,*

$$\|u_\epsilon - u\|_{L^2(0,T;W_0^{1,2}(D))} \to 0, \quad as\ \epsilon \to 0. \tag{2.86}$$

***Proof***

1. We start by establishing some a priori estimates that will be useful.
   Consider the weak form of the penalized problem, with a test function $v - u_\epsilon$, where $v \in K$ (arbitrary)

$$\int_0^T \langle \frac{\partial u_\epsilon}{\partial t}, v - u_\epsilon\rangle dt + \int_0^T \mathbf{a}(u_\epsilon, v - u_\epsilon)dt + \int_0^T \langle \beta_\epsilon(u_\epsilon), v - u_\epsilon\rangle dt - \int_0^T \langle f, v - u_\epsilon\rangle dt = 0.$$

Subtracting the term $\beta_\epsilon(v) = 0$, expressing $\mathbf{a}(u_\epsilon, v - u_\epsilon) = -\mathbf{a}(v - u_\epsilon, v - u_\epsilon) - \mathbf{a}(v, v - u_\epsilon)$, and $\langle \frac{\partial u_\epsilon}{\partial t}, v - u_\epsilon\rangle = -\langle \frac{\partial (v-u_\epsilon)}{\partial t}, v - u_\epsilon\rangle + \langle \frac{\partial v}{\partial t}, v - u_\epsilon\rangle$, we obtain

$$\begin{aligned}
&-\int_0^T \langle \frac{\partial (v - u_\epsilon)}{\partial t}, v - u_\epsilon\rangle dt - \int_0^T \mathbf{a}(v - u_\epsilon, v - u_\epsilon)dt \\
&-\int_0^T \langle \beta_\epsilon(v) - \beta_\epsilon(u_\epsilon), v - u_\epsilon\rangle dt - \int_0^T \langle f, v - u_\epsilon\rangle dt \\
&+\int_0^T \mathbf{a}(v, v - u_\epsilon)dt + \int_0^T \langle \frac{\partial v}{\partial t}, v - u_\epsilon\rangle dt = 0,
\end{aligned}$$

which upon integrating the first term yields

$$-\frac{1}{2}\underbrace{\|v-u_\epsilon\|^2_{L^2(D)}(T)}_{\geq 0}+\frac{1}{2}\|v-u_\epsilon\|^2_{L^2(D)}(0)-\int_0^T\underbrace{\mathbf{a}(v-u_\epsilon,v-u_\epsilon)}_{\geq 0}dt$$
$$-\int_0^T\underbrace{\langle\beta_\epsilon(v)-\beta_\epsilon(u_\epsilon),v-u_\epsilon\rangle}_{\geq 0}dt$$
$$-\int_0^T\langle f,v-u_\epsilon\rangle dt+\int_0^T\mathbf{a}(v,v-u_\epsilon)dt+\int_0^T\langle\frac{\partial v}{\partial t},v-u_\epsilon\rangle dt=0, \tag{2.87}$$

and we observe that (conveniently!) the underbraced terms keep their sign, whereas the second term only depends on the data since $u_\epsilon(0)=u_0$.

Using (2.87) as a starting point, we may obtain various useful estimates. For example, solving with respect to the term involving $\mathbf{a}(v-u_\epsilon,v-u_\epsilon)$, we obtain

$$\int_0^T\mathbf{a}(v-u_\epsilon,v-u_\epsilon)dt\leq-\int_0^T\langle f,v-u_\epsilon\rangle dt+\int_0^T\mathbf{a}(v,v-u_\epsilon)dt$$
$$+\int_0^T\langle\frac{\partial v}{\partial t},v-u_\epsilon\rangle dt+\frac{1}{2}\|v(0)-u_0\|^2_{L^2(D)}, \tag{2.88}$$

where the inequality comes from the terms in (2.87) that we omitted as we have information on their sign. We now estimate further (2.88) using coercivity of $\mathbf{a}$ for the LHS and then for the RHS (a) continuity for $\mathbf{a}$, (b) the Cauchy-Schwarz inequality for all terms, and (c) the Poincaré inequality to pass from the $L^2(D)$ norms to the $W_0^{1,2}(D)$ norms for the second and third to obtain

$$c_0\|v-u_\epsilon\|^2_{L^2(0,T;W_0^{1,2}(D))}\leq c_P\|f\|_{L^2(0,T;L^2(D))}\|v-u_\epsilon\|_{L^2(0,T;W_0^{1,2}(D))}$$
$$+c_1\|v\|_{L^2(0,T;L^2(D))}\|v-u_\epsilon\|_{L^2(0,T;W_0^{1,2}(D))}$$
$$+c_P\|\frac{\partial v}{\partial t}\|_{L^2(0,T;L^2(D))}\|v-u_\epsilon\|_{L^2(0,T;W_0^{1,2}(D))}$$
$$+\frac{1}{2}\|v(0)-u_0\|^2_{L^2(D)}, \tag{2.89}$$

which leads to an estimate

$$\|v-u_\epsilon\|_{L^2(0,T;W_0^{1,2}(D))}\leq c \implies \|u_\epsilon\|_{L^2(0,T;W_0^{1,2}(D))}\leq c, \tag{2.90}$$

for a constant $c$ depending only on the choice of $v$, $f$ and the initial data and importantly independent of $\epsilon$. For the second estimate we simply used the reverse triangle inequality and the fact that $v$ is arbitrary.

We may also solve (2.87) with respect to the term involving $\beta_\epsilon$, to obtain

$$0 \le \underbrace{\int_0^T \langle \beta_\epsilon(v) - \beta_\epsilon(u_\epsilon), v - u_\epsilon\rangle \, dt}_{\ge 0} \le -\int_0^T \langle f, v - u_\epsilon\rangle dt + \int_0^T \mathbf{a}(v, v - u_\epsilon) dt$$
$$+ \int_0^T \langle \frac{\partial v}{\partial t}, v - u_\epsilon\rangle dt + \frac{1}{2}\|v(0) - u_0\|^2_{L^2(D)}, \tag{2.91}$$

where again the second inequality comes by omitting terms that keep their sign, while the first one comes from monotonicity of the penalty function. Estimating the RHS as above and using (2.90), along with the definition of the penalty function, we obtain the estimate

$$0 \le \int_0^T \langle \beta(v) - \beta(u_\epsilon), v - u_\epsilon\rangle dt \le c\,\epsilon, \tag{2.92}$$

for a constant $c$ independent of $\epsilon$, which conveys important information as to the behaviour of the penalty function and ultimately the solution as $\epsilon \to 0$ (as in Proposition 2.76).

2. We now need to obtain a priori bounds on $\frac{\partial u_\epsilon}{\partial t}$. For this we require a bound on $\beta_\epsilon(u_\epsilon)$ in $L^2(0, T; L^2(D))$ that is independent of $\epsilon$ (this is not guaranteed by step 1). While there are other ways around this calculation, here to simplify the argument we make a regularity assumption on the obstacle, i.e., that $\frac{\partial \psi}{\partial t}$ and $A\psi$ are in $L^2(D)$.

   In order to achieve such a bound, we work again as in Proposition 2.77 and use $\beta(u_\epsilon)$ as test function in the weak form for the penalized problem (2.85) so as to obtain an estimate for $\|\beta_\epsilon(u_\epsilon)\|_{L^2(0,T;L^2(D))}$ which does not depend on $\epsilon$. Note that by the properties of $u_\epsilon$ and $\beta$, $v = \beta_\epsilon(u_\epsilon)$ can indeed qualify as a test function in (2.85). This yields

$$\int_0^T \langle \frac{\partial u_\epsilon}{\partial t}, \beta_\epsilon(u_\epsilon)\rangle dt + \int_0^T \mathbf{a}(u_\epsilon, \beta_\epsilon(u_\epsilon)) dt + \underbrace{\int_0^T \langle \beta_\epsilon(u_\epsilon), \beta_\epsilon(u_\epsilon)\rangle dt}_{=\|\beta_\epsilon(u_\epsilon)\|_{L^2(0,T,L^2(D))}}$$
$$= \int_0^T \langle f, \beta_\epsilon(u_\epsilon)\rangle dt. \tag{2.93}$$

All the other terms except the first are exactly as those of the elliptic case, under an integral sign. We focus on the first term that (see Example 2.74) is expressed as

$$\int_0^T \langle \frac{\partial u_\epsilon}{\partial t}, \beta_\epsilon(u_\epsilon)\rangle dt = \frac{1}{\epsilon}\int_0^T \langle \frac{\partial (u_\epsilon - \psi)}{\partial t}, -(u_\epsilon - \psi)^-\rangle dt$$
$$+\frac{1}{\epsilon}\int_0^T \langle \frac{\partial \psi}{\partial t}, -(u_\epsilon - \psi)^-\rangle dt$$
$$= \frac{1}{\epsilon}\int_0^T \langle \frac{\partial (u_\epsilon - \psi)^-}{\partial t}, (u_\epsilon - \psi)^-\rangle dt - \frac{1}{\epsilon}\int_0^T \langle \frac{\partial \psi}{\partial t}, (u_\epsilon - \psi)^-\rangle dt$$
$$= \frac{1}{2\epsilon}\|(u_\epsilon - \psi)^-(T)\|^2_{L^2(D)} - \frac{1}{\epsilon}\int_0^T \langle \frac{\partial \psi}{\partial t}, (u_\epsilon - \psi)^-\rangle dt$$
$$= \frac{\epsilon}{2}\|\beta_\epsilon(u_\epsilon(T))\|^2_{L^2(D)} + \int_0^T \langle \frac{\partial \psi}{\partial t}, \beta_\epsilon(u_\epsilon)\rangle dt. \tag{2.94}$$

Keeping only the term involving $\|\beta_\epsilon(u_\epsilon)\|^2_{L^2(D)}$ on the LHS of (2.93), moving all other terms to the RHS, and using estimate (2.94) for the term involving $\frac{\partial u_\epsilon}{\partial t}$ and the estimate (2.77) (obtained in Proposition 2.77) for the term involving **a**, we obtain

$$\|\beta_\epsilon(u_\epsilon)\|^2_{L^2(0,T;L^2(D))} := \int_0^T \|\beta_\epsilon(u_\epsilon)\|^2_{L^2(D)} dt \leq \int_0^T \langle f - A\psi - \frac{\partial \psi}{\partial t}, \beta_\epsilon(u_\epsilon)\rangle dt$$
$$-\frac{\epsilon}{2}\|\beta_\epsilon(u_\epsilon(T))\|^2_{L^2(D)}\rangle dt$$
$$\leq \int_0^T \|f - A\psi - \frac{\partial \psi}{\partial t}\|_{L^2(D)}\|\beta_\epsilon(u_\epsilon)\|_{L^2(D)} dt$$
$$\leq \left(\int_0^T \|f - A\psi - \frac{\partial \psi}{\partial t}\|^2_{L^2(D)}\right)^{1/2}\left(\int_0^T \|\beta_\epsilon(u_\epsilon)\|^2_{L^2(D)} dt\right)^{1/2}, \tag{2.95}$$

from which, dividing both sides with $\|\beta_\epsilon(u_\epsilon)\|_{L^2(0,T;L^2(D))}$, we obtain the desired estimate

$$\|\beta_\epsilon(u_\epsilon)\|_{L^2(0,T;L^2(D))} \leq \|f - A\psi - \frac{\partial \psi}{\partial t}\|_{L^2(0,T;L^2(D))}$$
$$\leq \|f\|_{L^2(0,T;L^2(D))} + \|A\psi\|_{L^2(0,T;L^2(D))} + \|dt\psi\|_{L^2(0,T;L^2(D))}. \tag{2.96}$$

Note that we can still obtain a bound like (2.96) in terms of $\|f\|_{L^2(0,T;(W_0^{1,2}(D))')}$.

We are now ready for the bound of $\frac{\partial u_\epsilon}{\partial t}$. We take the weak form of the penalized problem for any test function $v \in L^2(0,T;W_0^{1,2}(D))$ such that

$\|v\|_{L^2(0,T;W_0^{1,2}(D))} \leq 1$ and observe that this yields

$$|\int_0^T \langle \frac{\partial u_\epsilon}{\partial t}, v \rangle dt| = | - \int_0^T \mathbf{a}(u_\epsilon, v)dt - \int_0^T \langle \beta_\epsilon(u_\epsilon), v\rangle dt + \int_0^T \langle f, v\rangle dt|$$

$$\leq c_1 \|u_\epsilon\|_{L^2(0,T;W_0^{1,2}(D))} \|v\|_{L^2(0,T;W_0^{1,2}(D))}$$

$$+ \|\beta_\epsilon(u_\epsilon)\|_{L^2(0,T;L^2(D))} \|v\|_{L^2(0,T;L^2(D))}$$

$$+ \|f\|_{L^2(0,T;(W_0^{1,2}(D))')} \|v\|_{L^2(0,T;W_0^{1,2}(D))}$$

$$\leq c_1 \|u_\epsilon\|_{L^2(0,T;W_0^{1,2}(D))} \|v\|_{L^2(0,T;W_0^{1,2}(D))}$$

$$+ c_P \|\beta_\epsilon(u_\epsilon)\|_{L^2(0,T;L^2(D))} \|v\|_{L^2(0,T;W_0^{1,2}(D))}$$

$$+ \|f\|_{L^2(0,T;(W_0^{1,2}(D))')} \|v\|_{L^2(0,T;W_0^{1,2}(D))}$$

$$\leq c_1 \|u_\epsilon\|_{L^2(0,T;W_0^{1,2}(D))} \| + c_P \|\beta_\epsilon(u_\epsilon)\|_{L^2(0,T;L^2(D))}$$

$$+ \|f\|_{L^2(0,T;(W_0^{1,2}(D))')},$$

where we used continuity of **a**, the Cauchy-Schwarz inequality, and the Poincaré inequality. Recalling the definition of the norm of the dual space, we can see that (see Definition 2.4)

$$\|\frac{\partial u_\epsilon}{\partial t}\|_{L^2(0,T;(W_0^{1,2}(D))')} \leq c_1 \|u_\epsilon\|_{L^2(0,T;W_0^{1,2}(D))} \| + c_P \|\beta_\epsilon(u_\epsilon)\|_{L^2(0,T;L^2(D))}$$
$$+ \|f\|_{L^2(0,T;(W_0^{1,2}(D))')},$$

where by (2.90) and (2.96) the RHS is bounded by a term which is independent of $\epsilon$. Hence, there exists some $c$ independent of $\epsilon$, such that

$$\|\frac{\partial u_\epsilon}{\partial t}\|_{L^2(0,T;(W_0^{1,2}(D))')} \leq c. \tag{2.97}$$

$2\frac{1}{2}$. The a priori bounds of 1 and 2 can be used to show the existence of solution for the penalized problem using either a Galerkin type approach (where arguments similar to step 3 below must be employed) or a fixed point scheme based on the solvability of the linear system (see, e.g., Evans (2010) Theorem 3, Section 7.1, or Theorem 11.7 Chipot (2000)), the a priori bounds for the non-linear term, and a fixed point theorem. The details are left to the reader for lack of sufficient space.

3. By the results of steps 1 and 2 and the uniform bounds obtained for $u_\epsilon$ in $L^2(0,T; W_0^{1,2}(D))$ and $\frac{\partial u_\epsilon}{\partial t}$ in $L^2(0,T; (W_0^{1,2}(D))')$ (see estimates (2.90) and (2.97)) imply the existence of a $u \in L^2(0,T; W_0^{1,2}(D) \cap W^{2,2}(D))$ with $\frac{\partial u}{\partial t} \in L^2(0,T; L^2(D))$ such that

$$u_\epsilon \rightharpoonup u, \text{ in } L^2(0,T; W_0^{1,2}(D)), \text{ and } \frac{\partial u_\epsilon}{\partial t} \rightharpoonup \frac{\partial u}{\partial t} \text{ in } L^2(0,T; L^2(D))$$

(up to subsequences) with all convergences being weak. By (2.83) we have that $u \in C([0,T]; L^2(D))$. Moreover, the Aubin-Lions lemma (Theorem 2.85) implies that the convergence $u_\epsilon \to u$ is strong in $L^2(0,T; L^2(D))$ (up to subsequences). Finally (2.95) yields that $\|\beta(u_\epsilon)\|_{L^2(0,T;L^2(D))} \le c\,\epsilon$, where $c$ is a constant depending on the data (independent of $\epsilon$).

Passing to the limit as $\epsilon \to 0$ in (2.91) (if the obstacle is not sufficiently smooth) or in (2.96) (for sufficiently smooth obstacle) and using the Lipschitz continuity of $\beta$ and the strong convergence $u_\epsilon \to u$ in $L^2(D)$, we obtain that $\beta(u) = 0$; hence $u \in K$.

Next we return to estimate (2.88) for $v = u \in K$ and note that in the limit as $\epsilon \to 0$ the RHS of (2.88) tends to 0 (by the above weak convergence results); hence using also the coercivity of $\mathbf{a}$, we obtain from (2.88) that

$$0 \le \limsup_{\epsilon \to 0^+} c_0 \|u - u_\epsilon\|^2_{L^2(0,T;W_0^{1,2}(D))} \le \limsup_{\epsilon \to 0^+} \int_0^T \mathbf{a}(u - u_\epsilon, u - u_\epsilon) dt \le 0,$$

which leads to the result that

$$\lim_{\epsilon \to 0^+} \|u - u_\epsilon\|_{L^2(0,T;W_0^{1,2}(D))} = 0, \tag{2.98}$$

which is the strong convergence $u_\epsilon \to u$ in $L^2(0,T; W_0^{1,2}(D))$.

4. To obtain convergence of $u_\epsilon$ to the solution of the VI, we work as in the case of the elliptic problem. Consider any $v \in K$, and use $v - u_\epsilon \in W_0^{1,2}(D)$ as the test function in the weak formulation of the penalized problem. This yields

$$\begin{aligned} \int_0^T \langle \frac{\partial u_\epsilon}{\partial t}, v - u_\epsilon \rangle dt + \int_0^T \mathbf{a}(u_\epsilon, v - u_\epsilon) dt - \int_0^T \langle f, v - u_\epsilon \rangle dt \\ = \int_0^T \underbrace{\langle \beta_\epsilon(v) - \beta_\epsilon(u_\epsilon), v - u_\epsilon \rangle}_{\ge 0, \text{ by } (2.68)} dt \ge 0. \end{aligned} \tag{2.99}$$

All terms on the LHS, but the first, are essentially the same as those encountered in the elliptic case, so in the limit as $\epsilon \to 0^+$ they converge to the corresponding terms with $u_\epsilon$ replaced by $u$ (we need the convergence $u_\epsilon \to u$

in $L^2(0, T; W_0^{1,2}(D)))$. The first term is rearranged as

$$\int_0^T \langle \frac{\partial u_\epsilon}{\partial t}, v - u_\epsilon \rangle dt = \int_0^T \langle \frac{\partial u_\epsilon}{\partial t}, v - u \rangle dt$$
$$+ \int_0^T \langle \frac{\partial u_\epsilon}{\partial t}, (u - u_\epsilon) \rangle dt \to \int_0^T \langle \frac{\partial u}{\partial t}, v - u \rangle dt,$$

where we used the fact that $\frac{\partial u_\epsilon}{\partial t} \rightharpoonup \frac{\partial u}{\partial t}$ in $L^2(0, T, (W_0^{1,2}(D))')$. The second term converges to 0 since

$$|\int_0^T \langle \frac{\partial u_\epsilon}{\partial t}, (u - u_\epsilon) dt| \leq \| \frac{\partial u_\epsilon}{\partial t} \|_{L^2(0,T;(W_0^{1,2}(D))')} \|u - u_\epsilon\|_{L^2(0,T;W_0^{1,2}(D))}$$
$$\leq c\|u - u_\epsilon\|_{L^2(0,T;W_0^{1,2}(D))} \to 0,$$

by the fact that $\|\frac{\partial u_\epsilon}{\partial t}\|_{L^2(0,T;(W_0^{1,2}(D))')}$ is uniformly bounded in $\epsilon$ and $u_\epsilon \to u$ in $L^2(0, T; W_0^{1,2}(D))$ (strong). Putting it all together, we establish that (2.99) in the limit as $\epsilon \to 0$ is the weak formulation of the VI for $u$, i.e., $u$ solves the VI.

□

### Regularity

**Proposition 2.89** *Under the assumptions that the coefficients of $A$ are sufficiently smooth, $u_0 \in W_0^{1,2}(D)$, $f \in L^2(0, T; L^2(D))$, and $\frac{\partial \psi}{\partial t}$, $A\psi \in L^2(0, T; L^2(D))$, there exists a solution to the VI (2.84), which enjoys better regularity conditions and satisfies the bound*

$$\max_{t\in[0,T]} \|u(t)\|_{W_0^{1,2}(D)} + \|u\|_{L^2(0,T;W^{2,2}(D))} + \|\frac{\partial u}{\partial t}\|_{L^2(0,T;L^2(D))}$$
$$\leq c\Big( \|f\|_{L^2(0,T;L^2(D))} + \|\frac{\partial \psi}{\partial t}\|_{L^2(0,T;L^2(D))} + \|A\psi\|_{L^2(0,T;L^2(D))} + \|u_0\|_{W_0^{1,2}(D)} \Big). \tag{2.100}$$

***Proof***

1. The proof is based on the existence and regularity theory for linear parabolic equations of the form

$$\frac{\partial w}{\partial t} + Aw = h, \quad w(0) = w_0. \tag{2.101}$$

If the coefficients of $A$ are smooth, $a_{ij}, b_i, c \in C^1(D)$, and $w_0 \in W_0^{1,2}(D)$, then (see, e.g., Theorem 5, Section 7.1 Evans (2010)) the solution is more regular, satisfying $w \in L^2(0, T; W^{2,2}(D)) \cap L^\infty(0, T; W_0^{1,2}(D))$,

$\frac{\partial w}{\partial t} \in L^2(0,T;L^2(D))$, and we have the estimate

$$\begin{aligned} \operatorname*{ess\,sup}_{t\in[0,T]}\|w(t)\|_{W_0^{1,2}(D)} + \|w\|_{L^2(0,T;W^{2,2}(D))} + \|\frac{\partial w}{\partial t}\|_{L^2(0,T;L^2(D))} \\ \le C(\|h\|_{L^2(0,T;L^2(D))} + \|w_0\|_{W_0^{1,2}(D)}). \end{aligned} \tag{2.102}$$

2. We now obtain an estimate such as (2.100) for $u_\epsilon$ with bounds independent of $\epsilon$. The strategy is essentially the same as in the proof of Proposition 2.77, i.e., express (2.85) as

$$\frac{\partial u_\epsilon}{\partial t} + Au_\epsilon = \underbrace{f - \beta_\epsilon(u_\epsilon)}_{=h}, \quad u_\epsilon(0) = u_0, \tag{2.103}$$

with $h := f - \beta_\epsilon(u_\epsilon) \in L^2(0,T;L^2(D))$, and apply the estimate (2.102) to (2.103) to obtain the desired bound in terms of $\|f - \beta_\epsilon(u_\epsilon)\|_{L^2(0,T;L^2(D))}$. What is required in this strategy is a bound on $\beta_\epsilon(u_\epsilon)$ in $L^2(0,T;L^2(D))$ that is independent of $\epsilon$. But this is exactly estimate (2.96), which we have obtained in step 2 in the proof of Proposition 2.88. Using this estimate we obtain that

$$\begin{aligned} \max_{t\in[0,T]}\|u_\epsilon(t)\|_{W_0^{1,2}(D)} + \|u_\epsilon\|_{L^2(0,T;W^{2,2}(D))} + \|\frac{\partial u_\epsilon}{\partial t}\|_{L^2(0,T;L^2(D))} \\ \le c\Big(\|f\|_{L^2(0,T;L^2(D))} + \|\frac{\partial \psi}{\partial t}\|_{L^2(0,T;L^2(D))} + \|A\psi\|_{L^2(0,T;L^2(D))} + \|u_0\|_{W_0^{1,2}(D)}\Big), \end{aligned} \tag{2.104}$$

for a constant $c$ independent of $\epsilon > 0$. The above argument has to be complemented with a fixed point scheme (since the RHS of (2.103) depends on $u_\epsilon$).
3. Estimate (2.104) shows that $(u_\epsilon)_{\epsilon>0}$ is uniformly bounded in $L^2(0,T;W^{2,2}(D))$, so again by weak compactness arguments there exists a $\bar{u} \in L^2(0,T;W^{2,2}(D))$ such that $u_\epsilon \rightharpoonup \bar{u}$ in $L^2(0,T;W^{2,2}(D))$ (up to subsequences). Combining that with the results of step 3 in the proof of Proposition 2.88, by uniqueness of limit arguments (similar to those employed in the elliptic case), we conclude that $\bar{u} = u$, with $u$ already identified as the solution of the VI. Hence $u \in L^2(0,T;W^{2,2}(D))$. By using the weak lower semicontinuity of the norm, we can pass to the limit in the estimate (2.104) and recover the estimate for the solution $u$.

□

### 2.7.4 Fully Worked Example: Real Options in Finite Horizon

The results of this section directly apply to the problem of real options in finite horizon (see Sect. 1.2.3). Assume a real option on a multidimensional cash flow which follows the multidimensional GBM model (1.45) with an intertemporal pay-off function $f \in L^2(0, T; L^2(D))$ and a smooth enough instantaneous pay-off function $\psi$, such that $\frac{\partial \psi}{\partial t}, A\psi \in L^2(0, T; L^2(D))$. We have already established the continuity and coercivity of the corresponding bilinear form **a**. Then, a straightforward application of Proposition 2.89 shows that the value function $u \in L^2(0, T, W^{2,2}(D))$ and satisfies an estimate such as (2.104). Using the Sobolev embeddings and arguments similar to those in Sect. 2.6.7, we may establish continuity of $u$ or even $C^1$ regularity, as well as the linear complementarity formulation of the VI. If $\psi$ does not satisfy the required smoothness properties, we may always use a smooth approximation. Details are left to the reader.

## 2.8 Viscosity Solutions

### 2.8.1 Introduction

An interesting notion in the theory of variational inequalities is the notion of viscosity solutions. Viscosity solutions are a notion of solutions for non-linear equations (including those related to variational inequalities) using only continuity and not requiring at all differentiability (as, e.g., the weak solutions that were discussed in the previous sections). Viscosity solutions for second order problems have been introduced by Crandall et al (1992) and were connected with optimal control problems and optimal stopping problems (see, e.g., Fleming and Soner (2006), Oksendal and Reikvam (1998), etc.). They find interesting applications in management science and finance, both from the analytic point of view and from the numerical point of view (see, e.g., Fleming and Zariphopoulou (1991), Tourin and Zariphopoulou (1997), Nikolopoulos and Yannacopoulos (2010), Ngo and Pham (2016), etc.). This brief introduction is based on Dragoni (2018), Silvestre (2015), Fernández-Real and Ros-Oton (2023), Calder (2018), Katzourakis (2014), and Evans (2010).

**Do Remember 2.90 (Viscosity Solutions)** *Viscosity solutions concern solutions of equations of the general form*

$$F(x, u, Du, D^2u) = 0, \tag{2.105}$$

(continued)

**Do Remember 2.90** (continued)
*where Du and $D^2u$ are the gradient and Hessian of u, respectively, and F is a given non-linear function. A viscosity solution u needs only be continuous! The parabolic case can also be considered allowing F to depend also on $u_t$.*

**Example 2.91** Variational inequalities in their complementarity formulation can be expressed as (2.105). Upon defining $F(x, u, Du, D^2u) := \min(Au - f, u - \psi)$, then (2.105) becomes the complementarity formulation of the elliptic VI discussed in Proposition 2.81, similarly for the parabolic case. ◁

## 2.8.2 Definitions and Properties

**Notation 2.92** *For any two matrices $P_1, P_2 \in \mathbb{R}^{N\times N}$, the relation $P_1 \preceq P_2$ implies that $P_2 - P_1$ is a positive semidefinite matrix (when $N = 1$, this relation reduces to the standard $\leq$ relation).*

**Definition 2.93 (Proper Functions)** A function $F : \mathbb{R}^N \times \mathbb{R} \times \mathbb{R}^N \times \mathbb{R}^{N\times N} \to \mathbb{R}$ is called proper if it is continuous and satisfies the following monotonicity properties:

(i) $F(x, u_1, p, P) \leq F(x, u_2, p, P)$ for every $u_1 \leq u_2$.
(ii) $F(x, u, p, P_1) \geq F(x, u, p, P_2)$ for every $P_1 \preceq P_2$.

**Example 2.94** The function

$$\begin{aligned} F(x, u, Du, D^2u) &:= \min(-\Delta u + \lambda u - f, u - \psi) \\ &= \min(-Tr(D^2u) + \lambda u - f, u - \psi) \end{aligned}$$

is proper. ◁

We now give a definition for viscosity solutions. For technical reasons, we need to introduce the sets of upper and lower semicontinuous functions $USC(D)$, $LSC(D)$, respectively.

**Do Remember 2.95** *A function is called upper semicontinuous at $x \in D$ if $\limsup_{z\to x} u(z) \leq u(x)$ and lower semicontinuous if $u(x) \leq \liminf_{z\to x} u(z)$.*

*The set of upper and lower semicontinuous functions is denoted by $USC(D)$ and $LSC(D)$, respectively. Clearly, $USC(D) \cap LSC(D) = C(D)$.*

*If appropriately bounded, USC functions achieve their maximum in compact sets, whereas LSC functions achieve their minimum in compact sets.*

(a) (b)

**Fig. 2.2** (**a**) A smooth test function $\phi$ touching $u$ from above: $u(x)-\phi(x) \leq u(x_0)-\phi(x_0)$ ($u-\phi$ has local maximum at $x_0$). (**b**) A smooth test function $\phi$ touching $u$ from below. (Figure courtesy of Dr. G. Domazakis)

**Definition 2.96 (Viscosity Solutions)** Consider equation (2.105) with $F$ proper.

(i) $\underline{u} \in USC(D)$ is a viscosity subsolution of (2.105) if at any $\underline{x}_0 \in D$ and for any $\underline{\phi} \in C^2(D)$ such that $\underline{u} - \underline{\phi}$ has a local maximum at $\underline{x}_0$ (see Fig. 2.2a)

$$F(\underline{x}_0, \underline{u}(\underline{x}_0), D\underline{\phi}(\underline{x}_0), D^2\underline{\phi}(\underline{x}_0))$$
$$= F(\underline{x}_0, \underline{\phi}(\underline{x}_0), D\underline{\phi}(\underline{x}_0), D^2\underline{\phi}(\underline{x}_0)) \leq 0,$$

where, by shifting $\underline{\phi}$ by a constant, we may assume $\underline{u}(\underline{x}_0) = \underline{\phi}(\underline{x}_0)$.

(ii) $\overline{u} \in LSC(D)$ is a viscosity supersolution of (2.105) if at any $\overline{x}_0 \in D$ and any $\overline{\phi} \in C^2(D)$ such that $\overline{u} - \overline{\phi}$ has a local minimum at $\overline{x}_0$ (see Fig. 2.2b)

$$F(\overline{x}_0, \overline{u}(\overline{x}_0), D\overline{\phi}(\overline{x}_0), D^2\overline{\phi}(\overline{x}_0))$$
$$= F(\overline{x}_0, \overline{\phi}(\overline{x}_0), D\overline{\phi}(\overline{x}_0), D^2\overline{\phi}(\overline{x}_0)) \geq 0,$$

where, by shifting $\overline{\phi}$ by a constant, we may assume $\underline{u}(\underline{x}_0) = \underline{\phi}(\underline{x}_0)$.

(iii) $u$ is a viscosity solution of (2.105) if it is both a viscosity subsolution and a viscosity solution.

**Remark 2.97** By appropriately modifying $\underline{\phi}$ locally (e.g., by adding a term proportional to $|x_0 - x|^4$), we may assume that the local maximum of $\underline{u} - \underline{\phi}$ is strict, similarly for the supersolution case. ⊕

The intuition behind viscosity subsolutions is that any suitable test function (see Fig. 2.2) is a subsolution of the equation, where now all derivatives act on the smooth test function;[23] similarly for supersolutions. The terminology comes from viscosity perturbations of first order non-linear equations of the form $F_0(x, u, Du) + \epsilon^2 \Delta u = 0$ (see Evans (2010)). For alternative (and equivalent) definitions in terms of the concept of sub- and superjets (which are essentially sets of suitable test functions for sub- and supersolutions respectively, see e.g. Calder (2018)).

The notion of viscosity solutions is compatible with that of classical solutions, as long as the viscosity solution is smooth enough.

**Proposition 2.98** *The following hold:*

*(i) A classical solution of* (2.105) *is also a viscosity solution.*
*(ii) A viscosity solution of* (2.105) *which is twice differentiable at $x \in D$ coincides with the classical solution at $x$.*

***Proof***

(i) Let $u \in C^2(D)$ be a classical solution of (2.105), so that

$$F(x, u(x), Du(x), D^2u(x)) = 0, \quad \forall\, x \in D. \tag{2.106}$$

For any $\underline{x} \in D$ and any suitable test function $\underline{\phi}$ for the subsolution property, $u(\underline{x}) = \underline{\phi}(\underline{x})$, $Du(\underline{x}) = D\underline{\phi}(\underline{x})$, $D^2u(\underline{x}) - D^2\underline{\phi}(\underline{x}) \preceq 0$, the last relation coming from the second order condition for the local maximum of the function $u - \underline{\phi}$ at $\underline{x}$. Since $F$ is proper and $D^2\underline{\phi}(\underline{x}) \succeq D^2u(\underline{x})$, we have that

$$F(\underline{x}, \underline{\phi}(\underline{x}), D\underline{\phi}(\underline{x}), D^2\underline{\phi}(\underline{x})) \le F(\underline{x}, u(\underline{x}), Du(\underline{x}), D^2u(\underline{x})) \stackrel{(2.106)}{=} 0; \tag{2.107}$$

hence $u$ is a (viscosity) subsolution. Working in a similar fashion, we can show that $u$ is also a viscosity supersolution of (2.105), and the result follows.

(ii) (Sketch) If the viscosity solution $u$ is $C^2(D)$, then the result follows easily by the observation that we can use $u$ as a test function for both the sub- and the supersolution property, at any point $x \in D$. Then upon combining the two inequalities for $F$ we get that (2.106) holds; hence $u$ is a classical solution.

The situation becomes more interesting if $u$ is twice differentiable at a point $x_0 \in D$ and not on the whole of $D$, in which case the above argument has to be localized. Then we can show the existence of a smooth function $\underline{\phi} \in C^2(D)$ such that $\underline{\phi}(x_0) = u(x_0)$, $D\underline{\phi}(x_0) = Du(x_0)$, $D^2\underline{\phi}(x_0) = D^2u(x_0)$ with $u - \underline{\phi}$ displaying a local maximum at $x$, which can be used as a test function for the subsolution property. The function $\underline{\phi}$ locally (in a ball $B(x_0, r)$ for a suitable

[23] As a motivation for the definition of viscosity subsolution, consider a smooth solution of $-u'' = 0$, by the fact that $u - \underline{\phi}$ has a local maximum at $\underline{x}_0$, $(u - \underline{\phi})''(\underline{x}_0) \le 0$, and hence $-\underline{\phi}''(\underline{x}_0) \le 0$.

$r > 0$) looks like

$$\underline{\phi}(x) = u(x_0)+\langle Du(x_0), x-x_0\rangle+\frac{1}{2}\langle x-x_0, D^2u(x_0)(x-x_0)\rangle+|x-x_0|^2 g(\|x\|),$$

where $g : (0,r) \to \mathbb{R}_+$ is a $C^2$ function. This function can be extended over the whole of $D$ by standard arguments. By the properties of the function $\underline{\phi}$, this yields

$$F(x_0, \underline{\phi}(x_0), D\underline{\phi}(x_0), D^2\underline{\phi}(x_0)) = F(x_0, u(x_0), Du(x_0), D^2u(x_0)) \leq 0. \tag{2.108}$$

Similarly, there exists a function $\overline{\phi}$, which smoothly touches $u$ from below up to the second order at $x$, and such that $u - \overline{\phi}$ displays a local minimum at $x$, which can be used as a test function for the supersolution property, leading to (2.108) with the opposite sign. Then, combining the two we conclude that $u$ is a classical solution at $x$. For more details, see, e.g., Calder (2018) or Katzourakis (2014).

□

**Properties of Viscosity Solutions**

A property of viscosity solutions (making them useful for numerics) is stability.

**Theorem 2.99 (Stability of Viscosity Solutions)** *Let $(u_n)_{n\in\mathbb{N}} \subset C(D)$ be a sequence of continuous viscosity solutions of* (2.105)*, with $F$ replaced by the sequence of continuous (in all variables) functions $(F_n)_{n\in\mathbb{N}}$. If $F_n \to F$, continuous, and $u_n \to u$, locally uniformly,*[24] *then $u$ is a viscosity solution of* (2.105).

***Proof*** Clearly $(u_n)_{n\in\mathbb{N}}$ is a family of subsolutions. We will show that $u$ is a subsolution.

1. Consider a point $x \in D$ and any $\phi \in C^2(D)$, such that $u - \phi$ attains a local maximum at $x$, with $u(x) - \phi(x) = 0$. By possibly modifying $\phi$ locally (e.g., adding a term $\gamma\|x - z|^4$ for $\gamma > 0$), the local maximum can be considered strict, i.e., we can find a ball, of sufficiently small radius, $B(x,r) \subset\subset D$ with the property

$$u(z) < \phi(z), \quad \forall\, z \in B(x,r), \quad z \neq x. \tag{2.109}$$

2. We claim that we can find a sequence $(x_n)_{n\in\mathbb{N}}$ such that $u_n - \phi$ has a local maximum at $x_n$ for $n$ sufficiently large.

---

[24] That is, uniformly for all $D_0 \subset\subset D$. Recall that uniform convergence guarantees continuity of $u$.

To prove this claim, we work as follows: For any $n \in \mathbb{N}$, since $u_n - \phi$ is a continuous function, by Weierstrass maximum theorem it has a maximum $x_n \in \overline{B}(x, r)$. The sequence $(x_n)_{n\in\mathbb{N}}$ thus obtained is the required sequence.

We show that $x_n \to x$. Suppose not. Then, there exists a subsequence $(x_{n_k})_{k\in\mathbb{N}}$ and $\delta > 0$ such that $|x_{n_k} - x| > \delta$ for all $k$. Since $\overline{B}(x, r)$ is compact, by the Bolzano-Weierstrass theorem there exists a further subsequence (still denoted by $(x_{n_k})_{k\in\mathbb{N}}$ for convenience) and $\hat{x} \in \overline{B}(x, r)$ such that $x_{n_k} \to \hat{x}$. By the properties of the original subsequence $(x_{n_k})_{k\in\mathbb{N}}$ and passing to the limit, we have that $|x - \hat{x}| > \delta$. Since, by definition, $x_{n_k}$ is the point of a local maximum for $u_{n_k} - \phi$, we have that

$$u_{n_k}(x_{n_k}) - \phi(x_{n_k}) \geq u_{n_k}(x) - \phi(x). \tag{2.110}$$

Since $u_n \to u$ uniformly, we pass to the limit as $k \to \infty$ in (2.110), and using the continuity properties of $\phi$, we obtain

$$u(\hat{x}) - \phi(\hat{x}) \geq u(x) - \phi(x) = 0, \tag{2.111}$$

with the last equality coming since $u - \phi$ attains its maximum at $x$.

Hence, $u(\hat{x}) \geq \phi(\hat{x})$. But from (2.109) this implies that $\hat{x} = x$, which contradicts $|x - \hat{x}| > \delta$. Therefore, $x_n \to x$, and since $x$ is the center of $B(x, r)$ for sufficiently large $n$, $x_n$ is in the interior of the ball $B(x, r)$.

3. Since $x_n$ is a local maximum of $u_n - \phi$ and $u_n$ is a viscosity subsolution of (2.105), it holds that

$$F_n(x_n, u_n(x_n), D\phi(x_n), D^2\phi(x_n)) \leq 0. \tag{2.112}$$

By the fact that $\phi \in C^2(D)$ and $u_n \to u$ uniformly, letting $n \to \infty$ in (2.112) we obtain

$$F(x, u(x), D\phi(x), D^2\phi(x)) \leq 0; \tag{2.113}$$

hence $u$ is a viscosity subsolution of (2.105).

4. Working in a similar fashion for the supersolutions, we conclude the proof.

□

### 2.8.3 *Existence of Viscosity Solutions*

We briefly present the Perron method for existence of viscosity solutions (following Dragoni (2018) and Calder (2018)). This is based on a construction by Perron (1923) of solutions to $-\Delta u = 0$ (harmonic functions) in terms of the supremum of the set of all subsolutions (subharmonic functions). This was later extended to solutions of non-linear equations including the Hamilton-Jacobi equation in, e.g., Ishii (1987) and Crandall et al (1992).

We need the following definitions:

**Definition 2.100 (Comparison Principle)** Equation (2.105) satisfies the comparison principle if for any viscosity sub- and supersolution, $\underline{u}, \overline{u}$, respectively, satisfying the same boundary condition, it holds that $\underline{u} \leq \overline{u}$.

Whether an equation satisfies the comparison principle or not can be checked depending on its nature (see, e.g., Calder (2018)). Moreover, the following concept will be needed (see, e.g., Dal Maso (1993)).

**Definition 2.101 (Upper/Lower Semicontinuous Envelopes)** For any function $u$, we define the upper and lower semicontinuous envelopes, respectively, pointwise by

$$u^*(x) = \limsup_{z \to x,\, z \in D} u(z) = \lim_{r \to 0^+} \sup_{z \in B(x,r)} u(z) = \inf\{g(x) \, : \, g \in USC(D),\, g \geq u\},$$

$$u_*(x) = \liminf_{z \to x,\, z \in D} u(z) = \lim_{r \to 0^+} \inf_{z \in B(x,r)} u(z) = \sup\{g(x) \, : \, g \in LSC(D),\, g \leq u\}.$$

By definition $u^*$ is the smallest USC function above $u$ ($u_*$ the largest LSC below $u$).

Clearly, $u_* \leq u \leq u^*$, and $u$ is continuous at $x$ if and only if $u_*(x) = u^*(x) = u(x)$.

For equations satisfying the comparison principle a viscosity solution can be constructed using the family of subsolutions.

**Theorem 2.102** *Assume that* (2.105) *satisfies the comparison principle, and suppose that there exist viscosity sub- and supersolutions* $\underline{u}$ *and* $\overline{u}$, *respectively, satisfying the same boundary condition. Let*

$$w(x) := \sup\{\underline{w}(x) \, : \, \underline{u} \leq \underline{w} \leq \overline{u}, \;\; \underline{w} \text{ viscosity subsolution}\}. \tag{2.114}$$

*Then, $w$ is a viscosity solution of* (2.105) *satisfying the same boundary condition.*

The proof of Theorem 2.102 (which is essentially due to Lions-Ishii-Crandall Crandall et al (1992) and has been generalized by many other authors) requires two lemmas, which we state and prove before passing to the proof of the main theorem.

**Lemma 2.103 (Constructing New Subsolutions)** *Let $\mathbb{S} \neq \emptyset$ be a family of viscosity subsolutions of* (2.105). *Define* $\underline{w}(x) = \sup\{\underline{v}(x) \, : \, \underline{v} \in \mathbb{S}\}$.

*Then, $\underline{w}^*$ is a viscosity subsolution of* (2.105).[25]

[25] Note that we take $\underline{w}^*$ to make sure that this is a USC function. Unlike the infimum, the pointwise supremum of a family of USC functions is not necessarily USC. As an example consider the USC family consisting of $f_n(x) = \mathbf{1}_{[1/n,\infty)}(x)$; the pointwise supremum is $f(x) = \mathbf{1}_{(0,\infty)}(x)$ that is not USC.

***Proof (Sketch)*** Take an arbitrary $x \in D$, and consider $\phi \in C^2(D)$ a suitable test function for viscosity subsolutions. By a suitable local modification $\phi$ if necessary (e.g., with a quartic term as in Theorem 2.99), $\underline{w}^* - \phi$ achieves the value 0, a strict local maximum at $x$, i.e., there exists a ball $B(x, r)$ such that

$$\underline{w}^*(z) - \phi(z) \leq -\rho(z; x), \quad z \in B(x, r), \tag{2.115}$$

with $\rho(z; x) \geq 0$, $\lim_{z \to x} \rho(z; x) = 0$, and achieving the value 0 only at $z = x$.

We must somehow connect $\underline{w}^*$ with $\underline{w}$. We recall that, by the properties of the limsup, for every $z \in B(x, r)$ there exists a sequence $(z_n)_{n \in \mathbb{N}}$, $z_n \to z$, such that $\underline{w}^*(z) = \limsup_{y \to z} \underline{w}(y) = \lim_n \underline{w}(z_n)$. We apply this for $x$, construct the corresponding sequence $x_n \to x$, with this property, i.e., $\underline{w}^*(x) = \lim_n \underline{w}(x_n)$, and use the definition of the supremum to characterize $\underline{w}(x_n)$ for every $n$. Since $\underline{w} = \sup\{\underline{v} \, : \, \underline{v} \in \mathbb{S}\}$ for every $\epsilon = 1/n$, $\underline{w} - \epsilon$ is not an upper bound of the set $\{\underline{v} \, : \, \underline{v} \in \mathbb{S}\}$. Hence, for every $n \in \mathbb{N}$, there exists $\underline{v}_n \in \mathbb{S}$ such that

$$\underline{w}(x_n) - \frac{1}{n} < \underline{v}_n(x_n) \leq \underline{w}(x_n). \tag{2.116}$$

Since $\underline{v}_n$ is a viscosity subsolution, it is USC; hence $\underline{v}_n - \phi$ is USC also. By applying the Weierstrass maximum theorem (see Do Remember 2.95), there exists some $\hat{x}_n \in \bar{B}(x, r)$ in which $\underline{v}_n - \phi$ achieves its maximum value.

We will show that $\hat{x}_n \to x$. To this end, apply (2.115) to $(\hat{x}_n)_{n \in \mathbb{N}}$, to get

$$\begin{aligned}\rho(\hat{x}_n; x) \leq \phi(\hat{x}_n) - \underline{w}^*(\hat{x}_n) &\overset{(\underline{w} \leq \underline{w}^*)}{\leq} \phi(\hat{x}_n) - \underline{w}(\hat{x}_n) \overset{(\underline{v}_n \leq \underline{w},\ (2.116))}{\leq} \phi(\hat{x}_n) - \underline{v}_n(\hat{x}_n) \\ &\overset{\hat{x}_n \in \arg\max(\underline{v}_n - \phi)}{\leq} \phi(x_n) - \underline{v}_n(x_n) \overset{(2.116)}{\leq} \phi(x_n) - \underline{w}(x_n) + \frac{1}{n},\end{aligned}$$

where upon taking the limit as $n \to \infty$, and considering the properties of the sequence $(x_n)_{n \in \mathbb{N}}$ and of the function $\rho(z; x)$, leads to the result $\hat{x}_n \to x$.

Since $\underline{v}_n$ is a subsolution and $\hat{x}_n$ is a local max of $\underline{v}_n - \phi$, it holds that

$$F(\hat{x}_n, \underline{v}_n(\hat{x}_n), D\phi(\hat{x}_n), D^2\phi(\hat{x}_n)) \leq 0.$$

If $\underline{v}_n(\hat{x}_n) \to \underline{w}^*(x)$, then passing to the limit as $n \to \infty$, and using the continuity of $F$, yields that $\underline{w}^*$ is a viscosity subsolution.

To show that $\underline{v}_n(\hat{x}_n) \to \underline{w}^*(x)$, we work as follows: Note that

$$\underline{v}_n(\hat{x}_n) - \phi(\hat{x}_n) \overset{\hat{x}_n \in \arg\max(\underline{v}_n - \phi)}{\geq} \underline{v}_n(x_n) - \phi(x_n) \overset{(2.116)}{\geq} \underline{w}(x_n) - \phi(x_n) - \frac{1}{n}. \tag{2.117}$$

Since $\hat{x}_n \to x$, $x_n \to x$, and $\phi$ is continuous, taking the liminf on both sides of (2.117) yields (after eliminating $\phi(x)$ on both sides)

$$\liminf_n \underline{v}_n(\hat{x}_n) \geq \lim_n \underline{w}(x_n) = \underline{w}^*(x). \tag{2.118}$$

On the other hand, by the definition of $\underline{w}$ as the supremum over the family of subsolutions $\mathbb{S}$, we have that $\underline{v}_n(\hat{x}_n) \leq \underline{w}(\hat{x}_n) \leq \underline{w}^*(\hat{x}_n)$, so that taking the limsup on both sides of this inequality and keeping in mind that $\underline{w}^*$ is USC, we have that

$$\limsup_n \underline{v}(\hat{x}_n) \leq \limsup_n \underline{w}^*(\hat{x}_n) \leq \underline{w}^*(x). \tag{2.119}$$

Combining (2.118) and (2.119) obtains $\underline{v}(\hat{x}_n) \to \underline{w}^*(x)$ and concludes. □

The second lemma we require (see e.g., Calder (2018)) is the following:

**Lemma 2.104** *Let $\underline{u} \leq \overline{u}$ be a viscosity sub- and supersolution of* (2.105) *and*

$$\mathbb{S} := \{\underline{v} \; : \; \underline{v} \text{ viscosity subsolution of } (2.105)\ \underline{u} \leq \underline{v} \leq \overline{u}\}.$$

*If for some $\underline{v} \in \mathbb{S}$, the LSC envelope $\underline{v}_*$ is not a viscosity supersolution of* (2.105), *then there exists a viscosity subsolution of* (2.105) *$\underline{w} \in \mathbb{S}$ strictly larger than $\underline{v}$.*

***Proof (Sketch)*** Assume that $\underline{v}_*$ is not a viscosity supersolution of (2.105). Then, (by the negation of the definition) there exist $x_0 \in D$ and a function $\phi \in C^2(\mathbb{R}^N)$ such that $\underline{v}_* - \phi \geq 0 = \underline{v}_*(x_0) - \phi(x_0)$ has a local minimum with

$$F(x_0, \underline{v}_*(x_0), D\phi(x_0), D^2\phi(x_0)) = F(x_0, \phi(x_0), D\phi(x_0), D^2\phi(x_0)) < -\rho, \tag{2.120}$$

for some $\rho > 0$ (i.e., the above quantity is strictly negative).

Clearly, considering also the properties of the LSC envelope, we have that $\underline{v}_*(x_0) = \phi(x_0) \leq \overline{u}(x_0)$, but the inequality is strict. Assume not. Then, since $\overline{u} - \phi \geq \underline{v}_* - \phi \geq 0$, we have that $\overline{u} - \phi$ also has a local minimum at $x_0$, and since $\overline{u}$ is a supersolution $F(x_0, \phi(x_0), D\phi(x_0), D^2\phi(x_0)) \geq 0$, which contradicts (2.120). Hence $\underline{v}_*(x_0) = \phi(x_0) < \overline{u}(x_0)$ so that setting $2\epsilon := \overline{u}(x_0) - \phi(x_0) > 0$, we have that $\phi(x_0) + 2\epsilon = \overline{u}(x_0)$. By the continuity of these functions, we can find an $r > 0$ such that

$$\phi(x) + \epsilon < \overline{u}(x), \quad \forall\ x \in B(x_0, r). \tag{2.121}$$

We now perturb $\phi$ locally to

$$\hat{\phi}(x) = \phi(x) + \eta\left(\left(\frac{r}{2}\right)^4 - |x - x_0|^4\right)$$

and note that for any $\delta > 0$ by choosing $\eta$ sufficiently small, we can guarantee that $|D^k\hat{\phi}(x) - D^k\phi(x_0)| < \delta$, for $k = 0, 1, 2$ (where $D^0\phi = \phi$). Combining that with the continuity of $F$, we see that for $r$ sufficiently small, (2.120) leads to the result that $F(x, \hat{\phi}(x), D\hat{\phi}(x), D^2\hat{\phi}(x)) \le 0$, in $B(x_0, r)$; hence $\hat{\phi}$ is a (smooth) viscosity subsolution. Moreover, again by appropriate choice of $\eta$, (2.121) guarantees that $\hat{\phi} \le \bar{u}$ in $B(x_0, r)$. Finally, note that if $x \notin B(x_0, \frac{r}{2})$, then $(\frac{r}{2})^4 - |x - x_0|^4 \le 0$, so that $\hat{\phi}(x) \le \phi(x) \le u(x)$ in $B(x_0, r) \setminus B(x_0, \frac{r}{2})$.

Hence, upon defining

$$\underline{w}(x) = \begin{cases} \max(\underline{v}(x), \hat{\phi}(x)) & \text{in} \quad B(x_0, r) \\ \underline{v}(x) & \text{in } D \setminus B(x_0, r), \end{cases}$$

we see by Lemma 2.103 that it is a viscosity subsolution, which by construction is in $\mathbb{S}$ and strictly above $\underline{v}$ at some point. To see this last claim, note that $\underline{w}(x) \ge \hat{\phi}(x)$ in $B(x_0, r)$, so that choosing a sequence $\bar{x}_n \to x_0$, such that $\lim_n \underline{v}(\bar{x}_n) = \underline{v}_*(x_0)$ (this exists by the definition of $\underline{v}_*(x) = \liminf_{z \to x} \underline{v}(z)$), noting that along this sequence $\underline{w}(\bar{x}_n) \ge \hat{\phi}(\bar{x}_n)$, and taking the liminf as $n \to \infty$, we obtain that

$$\liminf_n \underline{w}(\bar{x}_n) \ge \lim_n \hat{\phi}(\bar{x}_n) = \phi(x_0) + \delta = \underline{v}_*(x_0) + \delta = \lim_n \underline{v}(\bar{x}_n) + \delta,$$

for some $\delta > 0$ (where for the first equality we used the properties of $\hat{\phi}$). Hence, for $n = N_0$ sufficiently large, $\underline{w}(\bar{x}_{N_0}) > \underline{v}(\bar{x}_{N_0})$, and the proof is complete. □

***Proof of Theorem 2.102*** By Lemma (2.103) $w^*$ is a viscosity subsolution of (2.105) in $\mathbb{S}$ as the USC envelope of the supremum of a family of viscosity subsolutions. Then, by the properties of USC envelopes and the comparison principle $w \le w^* \le \bar{u}$. However, since $w^* \in \mathbb{S}$ and $w$ is defined as the pointwise supremum over $\mathbb{S}$, we see that $w^* \le w$. Combining that with the above fact, obtained by the comparison principle, we have that $w = w^*$; hence $w$ is USC and is a subsolution.

We now show that $w$ is a supersolution as well. Here we use Lemma 2.104. To this end we first show that $w_*$ is a supersolution. Assume not. Then, there exists a subsolution $\hat{u} \in \mathbb{S}$ strictly larger at some point than $w$, a fact that contradicts the maximality of $w$. Hence $w_*$ is a supersolution, and since we have already established that $w$ is a subsolution, by the comparison principle $w \le w_*$. On the other hand, by the properties of LSC envelopes, $w_* \le w$, and hence $w = w_*$ so that $w$ is LSC and is also a supersolution.

If Dirichlet boundary conditions are required, these can be shown to be satisfied using the comparison principle once more. □

### 2.8.4 Fully Worked Example: Viscosity Solutions of Optimal Stopping Problems

The value function of optimal stopping problems of the type introduced in Sect. 1.2.3 can be naturally understood as viscosity solutions of relevant VIs. We show this for the general case of an American basket or a real put option. For simplicity of exposition we will consider the infinite horizon case (1.42) and work in terms of the multidimensional GBM problem (1.45). To avoid unnecessary technicalities, we may focus on operators as in (1.47); however, the result is much more general than that. We also require some technical assumptions concerning the uniform integrability of the family $(\psi(X_{\mathfrak{s}}))_{\mathfrak{s}\in\mathscr{T}}$ over the set $\mathscr{T}$ of suitable stopping times (see Oksendal and Reikvam (1998) or Oksendal (2013), Ch. 10). To characterize the value function as a viscosity solution, we follow Oksendal and Reikvam (1998).

**Proposition 2.105** *Subject to technical conditions, if the value function of problem* (1.42) *$u$ is continuous, then it is the viscosity solution of* $\min(Au-f, u-\psi)=0$ *in the sense of Definition 2.96.*

***Proof (Sketch)*** For simplicity we will assume that $A = -L + rI$, where $L$ is the generator of the Itō process $(X_t)_{t\geq 0}$ (the value of the underlying), $f = 0$, and we use the notation $\mathbb{E}_x$ to emphasize that $X_0 = x$. The results can be extended for more general cases.
0. We recall two important results:

(a) Dynkin's lemma (see, e.g., Oksendal (2013)), which follows essentially by Itō's lemma, according to which if $\phi$ is a $C^2$ function, for any stopping time $\mathfrak{s}$,

$$\mathbb{E}_x[e^{-r\mathfrak{s}}\phi(X_{\mathfrak{s}})] = \phi(x) + \mathbb{E}_x[\int_0^{\mathfrak{s}} \underbrace{e^{-rt}(L\phi - r\phi)(X_t)}_{=-e^{-rt}A\phi(X_t)}\,dt]. \tag{2.122}$$

(b) The characterization of the process $(e^{-rt}u(X_t))_{t\geq 0}$ as a (super)martingale for any $(t \geq \mathfrak{s}^*)$ $t \leq \mathfrak{s}^*$, where $\mathfrak{s}^* = \inf\{t : u(X_t) = \psi(X_t)\}$ is the optimal stopping time, i.e., for any stopping time $\mathfrak{s}$ (subject to certain technical conditions), $u(x) \overset{(\geq)}{=} \mathbb{E}_x[e^{-r\mathfrak{s}}u(X_{\mathfrak{s}})]$, $\mathfrak{s} \overset{(>)}{\leq} \mathfrak{s}^*$.

1. $u$ is a viscosity subsolution: For any $x \in D$, choose a smooth test function $\phi$ such that $u - \phi \leq 0 = u(x) - \phi(x)$ on a ball $B(x,\rho) \subset D$. It suffices to consider $x$ such that $u(x) > \psi(x)$ (subsolution property trivially holds otherwise).

For such $x$, choose a stopping time $\mathfrak{s} \leq \mathfrak{s}^*$ small enough so that $X_t \in B(x,\rho)$ for $t \in [0,\mathfrak{s}]$. Using the martingale property for the discounted value process and then applying Dynkin's lemma for $\mathfrak{s}$ and for the test function $\phi \in C^2$, such that $u \leq \phi$,

$$u(x) = \mathbb{E}_x[e^{-r\mathfrak{s}}u(X_{\mathfrak{s}})] \overset{(u\leq\phi)}{\leq} \mathbb{E}_x[\phi(X_{\mathfrak{s}})] \overset{(2.122)}{=} \phi(x) + \mathbb{E}_x[\int_0^{\mathfrak{s}} e^{-rt}(-A\phi)(X_t)dt],$$

which leads to

$$\mathbb{E}_x[\int_0^{\mathfrak{s}} e^{-rt}(-A\phi)(X_t)dt] \geq 0, \quad \forall \mathfrak{s},$$

sufficiently small. Dividing by $\mathfrak{s}$ and passing to the limit $\mathfrak{s} \to 0^+$ lead to the inequality $-A\phi(x) \geq 0$. Since $u(x) - \phi(x) = 0$, we have that $0 \leq u(x) - \psi(x) = \phi(x) - \psi(x)$, hence,

$$\begin{cases} A\phi(x) \leq 0, \\ \phi(x) - \psi(x) \geq 0 \end{cases} \implies \min\{A\phi(x), \phi(x) - \psi(x)\} \leq 0;$$

hence $u$ is a viscosity subsolution of the VI.
2. $u$ is a viscosity supersolution: For any $x \in D$, choose a smooth test function $\phi$ such that $u - \phi \geq 0 = u(x) - \phi(x)$ on a ball $B(x, \rho) \subset D$. For any stopping time $\mathfrak{s}$, by the supermartingale property of $(e^{-rt}u(X_t))_{t\geq 0}$, $u \geq \phi$ and Dynkin's formula,

$$u(x) \geq \mathbb{E}_x[e^{-r\mathfrak{s}}u(X_{\mathfrak{s}})] \overset{(u\geq\phi)}{\geq} \mathbb{E}_x[\phi(X_{\mathfrak{s}})] \overset{(2.122)}{=} \phi(x) + \mathbb{E}_x[\int_0^{\mathfrak{s}} e^{-rt}(-A\phi)(X_t)dt];$$

hence working similarly as above (sending $\mathfrak{s} \to 0^+$), $-A\phi(x) \leq 0$. Since $u(x) - \phi(x) = 0$, we have that $0 \leq u(x) - \psi(x) = \phi(x) - \psi(x)$, hence,

$$\begin{cases} A\phi(x) \geq 0, \\ \phi(x) - \psi(x) \geq 0 \end{cases} \implies \min\{A\phi(x), \phi(x) - \psi(x)\} \geq 0;$$

hence $u$ is a viscosity supersolution of the VI. □

## Appendix: Fixed Point Theorems

**Theorem 2.106 (Brouwer Fixed Point Theorem)** *Let $C$ be a closed convex subset of a finite dimensional normed vector space. If $F : C \to C$ is a continuous map, then it has a fixed point.*

**Theorem 2.107 (Banach Fixed Point Theorem)** *Let $(X, d)$ be a complete metric space and $S \subset X$ be a closed set. If $F : S \to S$ is a contraction, i.e., a map such that there exists $\varrho < 1$ for which $d(F(x), F(z)) \leq \varrho d(x, z)$ for all $x, z \in X$, then, $F$ has a unique fixed point.*

# References

Aubin JP (1963) Un théoreme de compacité. CR Acad Sci Paris 256(24):5042–5044
Bensoussan A, Lions JL (2011) Applications of variational inequalities in stochastic control. Elsevier
Brezis H (2011) Functional analysis, Sobolev spaces and partial differential equations, vol 2. Springer
Calder J (2018) Lecture notes on viscosity solutions. Lecture notes
Chipot M (1984) Variational inequalities and flow in porous media. Appl Math Sci 52
Chipot M (2000) Elements of nonlinear analysis. Springer Science & Business Media
Cioranescu D, Donato P, Roque MP (2018) An introduction to second order partial differential equations: Classical and variational solutions. World Scientific
Crandall MG, Ishii H, Lions PL (1992) User's guide to viscosity solutions of second order partial differential equations. Bull Am Math Soc 27(1):1–67
Dal Maso G (1993) An Introduction to $\Gamma$-convergence, vol 8. Birkhäuser
Dragoni F (2018) Introduction to viscosity solutions for nonlinear PDEs. Lecture notes
Evans LC (2010) Partial differential equations, vol 19. American Mathematical Society
Fernández-Real X, Ros-Oton X (2023) Regularity theory for elliptic PDE. arXiv preprint arXiv:230101564
Fleming WH, Soner HM (2006) Controlled Markov processes and viscosity solutions, vol 25. Springer Science & Business Media
Fleming WH, Zariphopoulou T (1991) An optimal investment/consumption model with borrowing. Math Oper Res 16(4):802–822
Harper P, Pang J (1990) Finite-dimensional variational inequalities and non-linear complementarity problems: a survey of theory, algorithms and applications. Math Program B 48:161–220
Ishii H (1987) Perron's method for hamilton-jacobi equations. Duke Math J 55:369–384
Katzourakis N (2014) An introduction to viscosity solutions for fully nonlinear PDE with applications to calculus of variations in $L^\infty$. Springer
Lax PD, Milgram AN (1954) Parabolic equations, volume 33 of annals of mathematics studies
Lewy H, Stampacchia G (1970) On the smoothness of superharmonics which solve a minimum problem. J Analyse Math 23:227–236
Lions JL, Stampacchia G (1967) Variational inequalities. Commun Pure Appl Math 20(3):493–519
Ngo MM, Pham H (2016) Optimal switching for the pairs trading rule: A viscosity solutions approach. J Math Anal Appl 441(1):403–425
Nikolopoulos C, Yannacopoulos A (2010) A model for optimal stopping in advertisement. Nonlinear Anal Real World Appl 11(3):1229–1242
Noor MA (1993) Wiener-Hopf equations and variational inequalities. J Optim Theory Appl 79:197–206
Oksendal B (2013) Stochastic differential equations: an introduction with applications. Springer Science & Business Media
Oksendal B, Reikvam K (1998) Viscosity solutions of optimal stopping problems. Stochast Stochast Rep 62(3–4):285–301
Perron O (1923) Eine neue behandlung der ersten randwertaufgabe für $\delta$u= 0. Math Zeitschrift 18:42–54
Rodrigues JF (1987) Obstacle problems in mathematical physics. Elsevier
Roubicek T (2013) Nonlinear partial differential equations with applications, vol 153. Springer Science & Business Media
Shi P (1991) Equivalence of variational inequalities with Wiener-Hopf equations. Proc Am Math Soc 111(2):339–346
Silvestre L (2015) Viscosity solutions of elliptic equations. Lecture Notes available at the webpage of the author 302
Tourin A, Zariphopoulou T (1997) Viscosity solutions and numerical schemes for investment/consumption models with transaction costs. Numerical methods in finance, pp 245–269

# Chapter 3
# Numerics

*Wherever there is number, there is beauty.*

*Proclus of Mallus (Προκλος Μαλλωτης)*

*If you come from mathematics, as I do, you realize that there are many problems, even classical problems, which cannot be solved by computation alone.*

*Roger Penrose*

## 3.1 Introduction

In this chapter we focus on numerical solutions of VI problems, by providing a variety of numerical methods, based on the analytical study of VIs introduced in Chap. 2. Our intention is to show how the theory developed so far can turn into practical, trustworthy methods for the numerical analysis of the solutions of VIs. The question of numerical solution of VIs is a very actively studied one, with a large variety of methods (depending on the problem at hand) and a huge related literature, impossible to cover in the limited space available here. We therefore choose a small selection of classes of methods, related to projection and proximal methods and duality methods. Moreover, as the choice of programming language is largely a matter of personal taste, needs and contingencies as to the environment we may find ourselves in, we will not provide code listings of the methods presented here in a particular computer language. Instead, we will provide pseudo-code for the algorithms, presented in a clear way so that each reader can transfer it to the programming language of choice. Our focus will be to guide the reader towards the construction of a numerical method for the resolution of some VI and present the necessary tools that allow us to prove rigorously that the proposed approximation scheme converges to the solution of the original problem.

 
A. E. Tsekrekos, A. N. Yannacopoulos, *Variational Inequalities in Management Science and Finance*, EURO Advanced Tutorials on Operational Research,
https://doi.org/10.1007/978-3-031-74119-7_3

## 3.2 Warm-Up: Projection Methods (From Fixed Points to Iterative Schemes)

Projection methods play a prominent role not only in analytic work but also in the numerical study of variational inequalities. The starting point for the projection method is Proposition 2.20 according to which

$$x^* \in C \, : \, \langle F(x^*), x - x^* \rangle \geq 0, \;\; \forall x \in C \implies$$
$$x^* \text{ is a fixed point of the operator } T_\gamma := \Pi_C(I - \gamma F) : C \to C, \text{ for any } \gamma > 0. \tag{3.1}$$

Recall that our analytic approach to the problem was also based on this connection.

The projection algorithm for the solution of VIs starts from this observation and essentially focuses on algorithms for computing the fixed point of the operator $T_\gamma$ defined above, hence approximating a solution for the VI. A standard way of approximating fixed points is through iterative procedures, an approach that works well for contraction operators. This leads to the following iterative scheme, called the basic projection scheme (often referred to as the Goldstein-Levitin-Polyak scheme, Goldstein (1964), Levitin and Polyak (1966))

$$x_{n+1} = \Pi_C(x_n - \gamma F(x_n)), \tag{3.2}$$

for a suitably selected step $\gamma > 0$ which may be constant or varying (i.e., for a sequence $(\gamma_n)_{n\in\mathbb{N}}$). In the special case where $F = D\Phi$, (3.2) reduces to the projected gradient scheme for the minimization of $\Phi$ on $C$.

**Algorithm 3.1 (Basic Projection Algorithm)**

- Input starting point $x_0 \in C$, and $\gamma > 0$.
- Until a convergence criterion is met iterate

$$x_{n+1} = \Pi_C(x_n - \gamma F(x_n)).$$

- When iteration stops return $x_n$.

This is an explicit scheme since each step produces the next approximation $x_{n+1}$ explicitly. The Goldstein-Levitin-Polyak method, as with all projection methods, is efficient if (a) the step $\gamma$ is chosen such that $T_\gamma = \Pi_C(I - \gamma F)$ is a contraction operator and (b) the projection operator on the convex closed set $C$ is known explicitly. Luckily the latter holds for a number of important examples of $C$ which arise often in applications. The iterative scheme (3.2) can be shown to converge to a solution of the VI under Lipschitz continuity and monotonicity conditions for an appropriate choice of $\gamma$ or $(\gamma_n)_{n\in\mathbb{N}}$ (related to the Lipschitz constant and the strong monotonicity coefficient of $F$). The choice of $\gamma$ is important for the convergence of the method, since $\Pi_C$ is a non-expansive operator, and iterative schemes for non-expansive operators may not converge. We will return to this discussion in Sect. 3.3.1.2. However, at this point it suffices to mention that this requires us to introduce more sophisticated algorithms than Algorithm 3.1.

We will not consider the convergence of the projection algorithm right now. Rather, we will identify the projection algorithm as a particular class of a larger class of algorithms called proximal algorithms and present a general proof for this class of algorithms, which will also allow us to study the convergence of various important extensions of the projection algorithm, such as, for example, the backward or implicit projection algorithm,

$$x_{n+1} = \Pi_C(x_n - \gamma_n F(x_{n+1})), \tag{3.3}$$

or the extragradient algorithm (Korpelevich (1976)) which is in some sense a mixture between the explicit and the implicit methods, in the spirit of a predictor-corrector method

$$\begin{aligned} z_n &= \Pi_C(x_n - \gamma_n F(x_n)), \\ x_{n+1} &= \Pi_C(x_n - \gamma_n F(z_n)), \end{aligned} \tag{3.4}$$

for some $\gamma_n > 0$ (this may be changing in each iteration or be kept constant). These algorithms may converge faster or under fewer assumptions on the data than the standard projection algorithm. For a different approach in the convergence proofs for projection-based algorithms, see, e.g., He and Liao (2002), Korpelevich (1976), and the references therein.

## 3.3 Proximal Point Algorithms

Proximal point algorithms (PPAs) constitute a general class of algorithms for solving equations (or rather inclusions) of the form

$$0 \in M(x), \tag{3.5}$$

where $M : X \to 2^X$ is a (possibly multi-valued) maximal monotone operator, and $X$ is a vector space finite or infinite dimensional. For the sake of simplicity, and since most numerical methods reduce to finite dimensional problems, we will assume $X$ to be finite dimensional, and in particular $X = \mathbb{R}^d$, but keep the general notation $X$ to emphasize the wide applicability of the algorithms presented.

Problems of the form (3.5) are strongly connected with VIs. As seen in Proposition 2.20,

$$x \in C \ : \ \langle F(x), z - x \rangle \geq 0, \quad \forall \, z \in C \iff 0 \in F(x) + N_C(x); \tag{3.6}$$

hence, upon defining

$$M = F + N_C, \tag{3.7}$$

we see that the problem of solving the VI reduces to the problem of finding a zero for the operator $M$ defined in (3.7). If $M$ is maximal monotone, then this opens up the route for using the general class of proximal point methods in the study of numerical solution of VIs. The maximal monotonicity of $M$ follows if $F$ is assumed monotone and by the maximal monotonicity of $N_C$.

## *3.3.1 Interlude: Maximal Monotone, Averaged, and Resolvent Operators*

The theory of proximal algorithms, the associated splittings, and their convergence properties build upon some fundamental facts concerning the class of monotone operators that enjoy maximal monotonicity. As this theory is an extension of monotonicity, the reader at this point may wish to review the material on monotonicity in Sect. 2.3.2, in particular Definition 2.21. We only state these facts here and refer the reader to the Appendix for more details or for some of the relevant proofs.

### 3.3.1.1 Maximal Monotone Operators

We recall that a (possibly multi-valued) operator $M$ is called monotone if

$$\langle w_x - w_z, x - z \rangle \geq 0, \ \ \forall x, z \in X, \ \ w_x \in M(x), \ \ w_z \in M(z). \tag{3.8}$$

If $M$ is single-valued, then $w_x = M(x)$, $w_z = M(z)$. Moreover, $\langle \cdot, \cdot \rangle$ can either be the inner product of $X$ or the pairing between $X'$, $X$ depending on whether $M : X \to 2^X$ or $M : X \to 2^{X'}$.

**Definition 3.2 (Maximal Monotone Operators)** A monotone operator $M$ is called maximal if

$$\forall x, z \in X, \ \ w_x \in M(x) \text{ such that } \langle x - z, w_x - w \rangle \geq 0 \implies w \in M(z). \tag{3.9}$$

Put otherwise, if the operator is considered as a graph, it is not strictly contained in any other monotone operator on $X$.

**Notation 3.3** *To facilitate the notation, we will use $M(x)$ ($M(z)$) invariably for the single-valued and the multi-valued case. We will also use the notation $Fix(M)$ for the set of fixed points of $M$ and $Zer(M)$ for the set of zeros of $M$.*

Maximal monotone operators have some very nice properties such as the fact that if $M$ is maximal monotone, then $I + M$ is surjective, i.e., $R(I + M) = X$, or that if $M$ is maximal monotone, then so is $M^{-1}$. But first we need to see how we can recognize maximal monotone operators (see, e.g., Ch. 20, Bauschke and Combettes (2017)).

**Proposition 3.4** *Let $X$ be a Hilbert space and $A : X \to X$ be a single-valued operator.*

*(i) If $A$ is a positive linear operator, then it is maximal monotone.*
*(ii) If $A$ is monotone and continuous, then it is maximal monotone.*
*(iii) If $A$ is monotone and hemicontinuous (i.e., $\lim_{\alpha \to 0^+} \langle z, A(x + \alpha w) \rangle = \langle z, A(x) \rangle$, $\forall\, x, z, w \in X$), then $A$ is maximal monotone.*
*(iv) If $A$ is non-expansive (i.e., $\|A(x) - A(z)\| \leq \|x - z\|$, $\forall\, x, z \in X$), then $I + \alpha A$ is maximal monotone for all $\alpha \in [-1, 1]$.*

On the other hand, there are also many multi-valued operators of great interest that are maximal monotone.

**Example 3.5** The following are examples of maximal monotone operators:

(i) The subdifferential $\partial\phi : X \to 2^X$ of a proper convex lower semicontinuous function $\phi : X \to \mathbb{R} \cup \{+\infty\}$ (see Definition 2.14) is a maximal monotone operator.
(ii) Let $C \subset X$ be a closed convex subset of a Hilbert space. The normal cone $N_C : X \to 2^X$ is a maximal monotone operator (see Example 2.15).
(iii) As in (ii) above, the projection operator $\Pi_C : X \to C$ is a maximal monotone operator.

These examples are important for the construction of numerical schemes for VIs. $\lhd$

The following properties of maximal monotone operators are essential for what follows:

**Proposition 3.6 (Properties of Maximal Monotone Operators)** *Let $X$ be a Hilbert space and $M : X \to 2^X$ be a maximal monotone operator. Then:*

*(i) For every $x \in X$, $M(x)$ is closed and convex.*
*(ii) (Minty) $R(I + M) = X$, i.e., $I + M$ is surjective.*[1]

Some care must be taken with the property that $A(x)$ is closed when $X$ is infinite dimensional. In this case we must also allow for weak convergence and closedness properties of $A(x)$ with respect to this kind of convergence. We will not

[1] In fact Minty's theorem guarantees that a monotone operator is maximally monotone if and only if $R(M + I) = X$.

require these properties at this context, where all problems addressed as assumed to be finite dimensional (even when we approximate infinite dimensional problems). We refer the reader to, e.g., Bauschke and Combettes (2017) or Kravvaritis and Yannacopoulos (2020) for details in the context of both infinite dimensional Hilbert spaces and Banach spaces, respectively.

#### 3.3.1.2 Averaged Operators

Averaged operators are a general class of operators for which iterative schemes perform well. They are defined in terms of non-expansive operators. Recall that an operator $N : X \to X$ is called non-expansive if $\|N(x_1) - N(x_2)\| \leq \|x_1 - x_2\|$ for all $x_1, x_2 \in X$. An important example of non-expansive operators is projection operators (see Proposition 2.16).

**Definition 3.7 (Averaged Operators)** An operator $T : X \to X$ is called $\theta$-averaged if there exists a non-expansive operator $N : X \to X$ such that $T = (1-\theta)I + \theta N$, for some $\theta \in (0, 1)$.

Clearly, $\theta$-averaged operators are non-expansive. The following equivalent characterizations of $\theta$-averaged operators (see Prop. 4.35, Bauschke and Combettes (2017)) are useful.

**Proposition 3.8** *The following are equivalent:*

(i) $T$ *is* $\theta$*-averaged.*

(ii) $2T - I$ *is* $2\theta$*-averaged.*

(iii) $(1-\theta)\|(I-T)x - (I-T)y\|^2 \leq \theta(\|x-y\|^2 - \|T(x) - T(y)\|^2)$.

$$(iv)\ \langle x - y, T(x) - T(y)\rangle \geq \frac{1-2\theta}{2(1-\theta)}\|x-y\|^2 + \frac{1}{2(1-\theta)}\|T(x) - T(y)\|^2. \tag{3.10}$$

The last assertion in (3.10) implies that certain $\theta$-averaged operators enjoy monotonicity properties. In fact, $\theta$-averaged operators with $\theta \in (0, \frac{1}{2}]$ are maximal monotone. This follows by the monotonicity of $T$ (from the last assertion in (3.10)) and its continuity (see Proposition 3.4(ii)).

Clearly, the class of averaged operators is closed under convex combinations. Moreover, an important fact is that the class of averaged operators is closed under composition (for a proof see Prop. 4.44 in Bauschke and Combettes (2017) or Lem. 3.7.2 in Kravvaritis and Yannacopoulos (2020)).

**Proposition 3.9** *Assume that* $T_i : X \to X$ *are* $\theta_i$*-averaged operators,* $i = 1, 2$. *Then* $T = T_1 \circ T_2$ *is* $\theta$*-averaged with* $\theta = \frac{\theta_1+\theta_2-2\theta_1\theta_2}{1-\theta_1\theta_2}$.[2]

[2] The result holds true for any finite number of compositions by appropriately modifying $\theta$.

The reason for the introduction of averaged operators is practical. Recall (see Theorem 2.107) that if $A$ is a contraction, i.e., $\|A(x_1) - A(x_2)\| \leq \rho \|x_1 - x_2\|$, for $\rho < 1$, then not only does there exist a solution of the equation $x = A(x)$ (a fixed point of $A$), but it can also be approximated by the iterative scheme $x_{n+1} = A(x_n)$. This is not true for a non-expansive operator $N$! However, as we will see in Proposition 3.11, we may express the problem of finding fixed points of non-expansive operators in terms of a fixed point problem for averaged operators and moreover in doing so obtain an iterative scheme that allows us to recover the required fixed point.

The following concept (see, e.g., Bauschke and Combettes (2017)) is important for what follows:

**Do Remember 3.10 (Fejér Sequence)** *A sequence $(x_n)_{n\in\mathbb{N}} \subset X$ is called Fejér with respect to a nonempty set $\mathfrak{F} \subset X$ if*

$$\|x_{n+1} - x\| \leq \|x_n - x\|, \quad \forall\, x \in \mathfrak{F},\ \forall\, n \in \mathbb{N}. \tag{3.11}$$

*Fejér sequences are bounded, and $(\|x_n - x\|)_{n\in\mathbb{N}}$ converges for any $x \in \mathfrak{F}$.*

Whether a Fejér sequence $(x_n)_{n\in\mathbb{N}}$ converges to a point in $\mathfrak{F}$ or does not depend on the nature of $\mathfrak{F}$ and whether $dim(X) < \infty$ or $\infty$ (the convergence being possibly weak in the latter case).

The following two classical results, originally due to Krasnoselskii, Mann, and Groetch (see, e.g., Thm 5.15, Prop. 5.16, Bauschke and Combettes (2017), Thm 1, Ryu and Yin (2022), Thm 3.4.9, and Kravvaritis and Yannacopoulos (2020)), are fundamental to numerical algorithms.

**Proposition 3.11 (Krasnoselskii-Mann)** *Assume that $T$ is a $\theta$-averaged operator in the form $T = (1-\theta)I + \theta N$ for some $\theta \in (0, 1)$ and $N$ non-expansive. Then, the iterative scheme*

$$x_{n+1} = T(x_n) = x_n + \theta(N(x_n) - x_n) \tag{3.12}$$

*converges globally (i.e., for any $x_0$) and linearly to an $x \in Fix(T) = Fix(N)$, i.e.,*

$$\|T(x_n - x_n)\|^2 := \|x_{n+1} - x_n\|^2 \leq \frac{\theta}{(n+1)(1-\theta)} dist(x_0, Fix(T))^2. \tag{3.13}$$

*In the case where $X$ is infinite dimensional, the convergence must be understood in the weak sense (even though $N(x_n) - x_n \to 0$ or $T(x_n) - x_n \to 0$ strongly).*

*Note that* (3.12) *is equivalent to a relaxation scheme for the solution of* $x = N(x)$.

***Proof*** $Fix(T) = Fix(N)$ is immediate. The rest of the proof proceeds in four steps.

1. The sequence $(\|x_n - x^*\|)_{n\in\mathbb{N}}$ is decreasing for any $x^* \in Fix(N)$.
   Indeed, for any $x^* \in X$ (irrespective of whether $x^* \in Fix(N)$),

$$\|x_{n+1} - x^*\|^2 = \|(1-\theta)x_n + \theta N(x_n) - x^*\|^2 = \|(1-\theta)(x_n - x^*) + \theta(N(x_n) - x^*)\|^2. \quad (3.14)$$

   On the other hand, for any $x, z \in X$, note that (by direct computation)

$$\|(1-\theta)x + \theta z\|^2 = (1-\theta)\|x\|^2 + \theta\|z\|^2 - \theta(1-\theta)\|x - z\|^2.$$

   Moreover, since $N$ is non-expansive, for $x^* \in Fix(N)$,

$$\|N(x) - x^*\| = \|N(x) - N(x^*)\| \leq \|x - x^*\|.$$

   Using the above in (3.14), we obtain the estimate

$$\begin{aligned}\|x_{n+1} - x^*\|^2 &= (1-\theta)\|x_n - x^*\|^2 + \theta\|N(x_n) - x^*\|^2 - \theta(1-\theta)\|N(x_n) - x_n\|^2\\ &\leq (1-\theta)\|x_n - x^*\|^2 + \theta\|x_n - x^*\|^2 - \theta(1-\theta)\|N(x_n) - x_n\|^2\\ &= \|x_n - x^*\|^2 - \theta(1-\theta)\|N(x_n) - x_n\|^2.\end{aligned} \quad (3.15)$$

2. Estimate (3.15) can guarantee that $(x_n)_{n\in\mathbb{N}}$ is a Fejér sequence with respect to the set $Fix(N) = Fix(T)$, i.e., $\|x_{n+1} - x^*\|^2 \leq \|x_n - x^*\|^2$, for all $x^* \in Fix(N) = Fix(T)$, and the sequence $(\|x_n - x_{n-1}\|)_{n\in\mathbb{N}}$ is decreasing, since

$$\|x_{n+1} - x_n\| = \|T(x_n) - T(x_{n-1})\| \leq \|x_n - x_{n-1}\|,$$

   where we used the fact that since $T$ is $\theta$-averaged, it is also non-expansive.
3. We now show the convergence rate. We fix an $m \in \mathbb{N}$ and sum the inequality (3.15) from $n = 0$ to $m$. This yields

$$\|x_{m+1} - x^*\|^2 \leq \|x_0 - x^*\|^2 - \frac{1-\theta}{\theta}\sum_{n=0}^{m}\|T(x_n) - x_n\|^2,$$

which upon rearranging gives

$$\sum_{n=0}^{m} \|T(x_n) - x_n\|^2 := \sum_{n=0}^{m} \|x_{n+1} - x_n\|^2 \leq \frac{\theta}{1-\theta}\|x_0 - x^*\|^2 - \frac{\theta}{1-\theta}\|x_{m+1} - x^*\|^2. \tag{3.16}$$

But the sequence $(\|x_n - x_{n-1}\|)_{n\in\mathbb{N}}$ is decreasing, and hence $\|x_{n+1} - x_n\| \geq \|x_{m+1} - x_m\|$ for all $n \leq m$, which by (3.16) yields

$$(m+1)\|x_{m+1} - x_m\|^2 \leq \frac{\theta}{1-\theta}\|x_0 - x^*\|^2 - \frac{\theta}{1-\theta}\|x_{m+1} - x^*\|^2 \leq \frac{\theta}{1-\theta}\|x_0 - x^*\|^2,$$

which upon rearranging, then taking the infimum of the RHS over all $x^* \in Fix(T)$, and switching the notation from $m$ to $n$ yields (3.13). Note that (3.13) implies

$$\|T(x_n) - x_n\| \to 0 \iff T(x_n) - x_n \to 0.$$

4. We now show, by compactness, that $(x_n)_{n\in\mathbb{N}}$ converges to some $x^* \in Fix(T)$.

If $X$ is finite dimensional, the fact that $(\rho_n)_{n\in\mathbb{N}}$, with $\rho_n = \|x_n - x^*\|$, is decreasing for any $x^* \in Fix(T)$ guarantees that the sequence $(x_n)$ lies in the closed ball $B(x^*, R)$ for an appropriate $R$, which is a compact set in finite dimensional spaces. Then it has a subsequence $(x_{n_k})_{k\in\mathbb{N}}$ that converges to some $\bar{x} \in B(x^*, R)$. Since $F(x_n) - x_n \to 0$, the same holds for the chosen converging subsequence, i.e., $T(x_{n_k}) - x_{n_k} \to 0$ as $k \to \infty$, so that passing to this limit we obtain (using also the continuity of $T$) that $T(\bar{x}) = \bar{x}$; hence $\bar{x} \in Fix(T)$. By the Fejér property of $(x_n)_{n\in\mathbb{N}}$ with respect to $Fix(T)$, we have that $(\rho'_n)_{n\in\mathbb{N}}$ with $\rho' = \|x_n - \bar{x}\|$ is a decreasing sequence; hence the whole sequence $(x_n)_{n\in\mathbb{N}}$ converges to $\bar{x} \in Fix(T)$. If $X$ is infinite dimensional, the above compactness argument must be replaced by a weak compactness argument. □

This interesting result generalizes for variable step relaxation schemes as well.

**Proposition 3.12 (Krasnoselskii-Mann-Groetsch)**

*(i) Let $N : X \to X$ be non-expansive. If $\theta_n \in [0, 1]$, with $\sum_{n\in\mathbb{N}} \theta_n(1-\theta_n) = \infty$, then, the variable step relaxation scheme*

$$x_{n+1} = T_n(x_n) := x_n + \theta_n(N(x_n) - x_n) \tag{3.17}$$

*converges to a point $x^* \in Fix(N)$. Note that $T_n$ is $\theta_n$-averaged.*

*(ii) Let $T : X \to X$ be $\theta$-averaged. If $\theta_n \in [0, \frac{1}{\theta}]$ with $\sum_{n\in\mathbb{N}} \theta_n(1 - \theta\theta_n) = +\infty$, then, the variable step relaxation scheme*

$$x_{n+1} = x_n + \theta_n(T(x_n) - x_n) \tag{3.18}$$

*converges to a point $x^* \in Fix(T)$.*

*In the case where $X$ is infinite dimensional the convergence must be understood in the weak sense (even though $N(x_n) - x_n \to 0$ or $T(x_n) - x_n \to 0$ strongly).*

***Proof (Sketch)***

(i) The proof is almost identical to the proof of Proposition 3.11, with minimal changes. For example, Step 1 is almost identical, with (3.15) modified to

$$\|x_{n+1} - x^*\|^2 \leq \|x_n - x^*\|^2 - \theta_n(1 - \theta_n)\|N(x_n) - x_n\|^2. \tag{3.19}$$

Upon adding (3.19) for all $n \in \mathbb{N}$, we obtain that

$$\sum_{n\in\mathbb{N}} \theta_n(1 - \theta_n)\|N(x_n) - x_n\|^2 \leq \|x_0 - x^*\|^2 < \infty;$$

hence, since $\sum_{n\in\mathbb{N}} \theta_n(1 - \theta_n) = +\infty$, we obtain that $\|N(x_n) - x_n\| \to 0$.

(ii) It follows from (i) since $T = (1 - \theta)I + \theta N$ for $N$ non-expansive.

□

This proposition is extremely useful from the point of view of numerical approximation of fixed points and leads to the celebrated Krasnoselski-Mann fixed point iteration scheme. What makes this result even more exciting is the fact that if $N$ is non-expansive, then the fixed point scheme $x_{n+1} = N(x_n)$ is not necessarily convergent, in contrast to the fixed point scheme $x_{n+1} = ((1 - \theta)I + \theta N)(x_n)$, for some $\theta \in (0, 1)$, which converges to some element in $Fix(N)$.

#### 3.3.1.3 The Resolvent Operator

Three very important operators related to maximal monotone operators are the resolvent operator, the related Cayley reflector, and the Yosida (or Moreau-Yosida) approximation.

**Definition 3.13** Let $M : X \to 2^X$ be a (possibly multi-valued) maximal monotone operator. We define[3] the single-valued operators:

[3] Please note that this notation differs between texts (e.g., Bauschke and Combettes (2017) and use the notation $J_M$ for the resolvent operator and $R_M$ for the Cayley operator).

(i) $R_M : X \to X$, called the resolvent of $M$, defined by

$$R_M = (I + M)^{-1} \iff z = R_M(x) \text{ if and only if } x \in z + M(z).$$

(ii) $C_M : X \to X$, called the Cayley reflector operator, defined by $C_M := 2R_M - I$.
(iii) The family of single-valued operators $(Y_{\gamma,M} : X \to X)_{\gamma>0}$, called the (Moreau)-Yosida approximation of $M$, defined by

$$Y_{\gamma,M} = \frac{1}{\gamma}(I - R_{\gamma M}).$$

The existence of the resolvent operator $R_M$ for a maximal monotone operator follows from Minty's surjectivity result (see Proposition 3.6(ii)). The fact that $(I + M)^{-1}$ is single-valued follows by the monotonicity of $M$. Hence, the maximal monotonicity of $M$ is needed to guarantee the existence of the resolvent.

The following examples will be of use in what follows.

**Example 3.14 (Resolvent Operator of the Normal Cone)** Let $M = N_C$, the normal cone of a convex closed set $C \subset X$. Then,

$$R_{\gamma M} = \Pi_C, \quad \forall\, \gamma > 0.$$

Indeed, by definition, for any $\gamma > 0$,

$$z = R_{\gamma M}(x) \iff x \in z + \gamma N_C(z).$$

This in turn, recalling Proposition 2.16(iii), leads to

$$x \in z + \gamma N_C(z) \iff x - z \in \gamma N_C(z) = N_C(z) \iff z = \Pi_C(x),$$

which is our claim. ◁

**Example 3.15 (Resolvent Operator of a Subdifferential, Proximal Operator)** Let $\phi : X \to \mathbb{R} \cup \{+\infty\}$ be a proper convex lower semicontinuous function and $M = \partial\phi : X \to 2^X$, its subdifferential, defined by

$$\partial\phi(x) = \{z \in X \ : \ \langle z, y - x\rangle \geq \phi(y) - \phi(x) \ \forall\, y \in X\}.$$

Then, $R_M$ is called the proximal operator $prox_\phi$. By definition,

$$z = prox_\phi(x) \iff x \in z + \partial\phi(z) \iff 0 \in z - x + \partial\phi(z) \iff$$

$$z = \arg\max_{y \in X} \left(\frac{1}{2}\|x - y\|^2 + \phi(y)\right),$$

where the uniqueness comes from the fact that the function $y \mapsto \frac{1}{2}\|x-y\|^2+\phi(y)$ is strictly convex. Hence, in the case of the subdifferential the resolvent operator is defined in terms of the solution of a regularized optimization problem. ◁

**Example 3.16** It holds that

$$y = Y_{\gamma,M}(x) \iff x - \gamma y = R_{\gamma M}(x) \iff y \in M(x - \gamma y).$$

In this respect $Y_{\gamma,M}$ is an approximation of $M$. ◁

The following properties of $R_M$ and $C_M$ will be very useful in what follows.

**Proposition 3.17** *Let $M : X \to 2^X$ be a maximal monotone operator. The following hold:*

*(i)* *$C_M$ is non-expansive.*
*(ii)* *$R_M$ is $\frac{1}{2}$-averaged, hence firmly non-expansive and maximal monotone.*
*(iii)* *For every $\gamma > 0$, $Y_{\gamma,M}$ is maximal monotone and $\frac{1}{\gamma}$-Lipschitz continuous.*

***Proof***

(i) Consider any $x_i \in X$, and set $y_i = C_M(x_i)$, $i = 1, 2$. By definition, $y_i = C_M(x_i) = 2R_M(x_i) - x_i$; hence $\frac{1}{2}(y_i + x_i) = R_M(x_i)$, i.e.,

$$x_i \in \frac{1}{2}(x_i + y_i) + M\left(\frac{1}{2}(x_i + y_i)\right) \iff$$

$$\exists\ w_i \in M\left(\frac{1}{2}(x_i + y_i)\right), \text{ such that } \frac{1}{2}(x_i - y_i) = w_i \iff \frac{x_i + y_i}{2} = x_i - w_i. \tag{3.20}$$

By the monotonicity of $M$ we have that

$$\begin{aligned}\langle \frac{1}{2}(y_1 + x_1) - \frac{1}{2}(y_2 + x_2), w_1 - w_2\rangle \geq 0 \overset{(3.20)}{\iff} \\ \langle (x_1 - w_1) - (x_2 - w_2), w_1 - w_2\rangle \geq 0.\end{aligned} \tag{3.21}$$

We express $y_i$ in terms of $y_i = x_i - 2w_i$ and calculate

$$\begin{aligned}&\|y_1 - y_2\|^2 - \|x_1 - x_2\|^2 = \|x_1 - 2w_1 - (x_2 - 2w_2)\|^2 - \|x_1 - x_2\|^2 \\ &= 4(\|w_1 - w_2\|^2 - \langle x_1 - x_2, w_1 - w_2\rangle) = 4(\langle (w_1 - w_2) \\ &- (x_1 - x_2), w_1 - ew_2\rangle) \overset{(3.21)}{\leq} 0,\end{aligned}$$

from which follows that $C_M$ is non-expansive.

(ii) The assertion follows directly from (i), since by definition $R_M = \frac{1}{2}I + \frac{1}{2}C_M$.

(iii) Let $y_i = Y_{\gamma,M}(x_i) = \frac{1}{\gamma}(I - R_{\gamma M})(x_i)$. Then, $R_{\gamma M}(x_i) = x_i - \gamma y_i$; hence $x_i \in x_i - \gamma y_i + \gamma M(x_i - \gamma y_i)$, i.e.,

$$\exists\ w_i \in M(x_i - \gamma y_i) \text{ such that } x_i = x_i - \gamma y_i + \gamma w_i \implies w_i = y_i. \tag{3.22}$$

By the monotonicity of $M$,

$$0 \leq \langle x_1 - \gamma y_1 - (x_2 - \gamma y_2), w_1 - w_2 \rangle \overset{(3.22)}{=} \langle x_1 - \gamma y_1 - (x_2 - \gamma y_2), y_1 - y_2 \rangle \implies$$
$$\gamma \| y_1 - y_2 \|^2 \leq \langle x_1 - x_2, y_1 - y_2 \rangle,$$

from which estimating the RHS in terms of the Cauchy-Schwarz inequality and dividing through by $\|y_1 - y_2\|$ yield the stated property. □

Finally, as follows by direct computation, an important identity connecting $R_{\gamma M}$ and $C_{\gamma M}$ is that for any $\gamma > 0$,

$$C_{\gamma M} = (I - \gamma M)(I + \gamma M)^{-1} = (I - \gamma M) R_{\gamma M} \iff C_{\gamma M}(I + \gamma M) = (I - \gamma M). \tag{3.23}$$

### 3.3.2 *The Fundamental Proximal Point Algorithm*

The fundamental version of the proximal point algorithm is the iterative scheme

$$0 \in M(x_{n+1}) + \frac{1}{\gamma}(x_{n+1} - x_n) \iff x_{n+1} = R_{\gamma M}(x_n), \quad \gamma > 0. \tag{3.24}$$

Clearly, a fixed point for (3.24) satisfies $0 \in M(x)$.

Note that (3.24) is an implicit scheme, in the sense that in order to go from step $n$ to step $n + 1$ we still have an operator equation to solve, to obtain the next approximation $x_{n+1}$ in terms of the current approximation $x_n$. Note also the similarities with the implicit (sub)gradient scheme for optimization if $M = \partial \Phi$.

**Do Remember 3.18 (Proximal Point Algorithm)** *$M$ maximal monotone, $\gamma > 0$*

- *The proximal point algorithm is the iteration*

$$x_{n+1} = R_{\gamma M}(x_n), \tag{3.25}$$

(continued)

**Do Remember 3.18** (continued)

*for the $\frac{1}{2}$-averaged operator $R_{\gamma M}$, so by Krasnoselskii-Mann Proposition 3.11) it converges.*

- *Often we encounter the relaxation version: For $\gamma > 0$, $\lambda > 0$,*

$$x_{n+1} = x_n + \lambda(R_{\gamma M}(x_n) - x_n), \tag{3.26}$$

*whose convergence can be treated by Proposition 3.12.*

- *Variable step versions, in which $\gamma = \gamma_n$ or $\lambda_n$, are possible and can be treated in the spirit of Proposition 3.12).*

The proximal algorithm has nice convergence properties as for example the following proposition indicates.

**Proposition 3.19** *The proximal point scheme* (3.25) *converges to an $x \in Fix(M)$ for every initial condition $x_0$ and any $\gamma > 0$. If $dim(X) = \infty$, the convergence is in the weak sense, unless M is strongly monotone, in which case the convergence is strong.*

***Proof*** The convergence of the proximal point method follows from the fact that $R_{\gamma M}$ is an $\frac{1}{2}$-averaged operator (see Proposition 3.17) and the convergence properties of the Krasnoselskii-Mann iterative scheme (see Proposition 3.11). □

The convergence properties of variable step schemes can be studied either by the Krasnoselskii-Mann-Groetch results, in the case where $\gamma_n = \gamma$ and the relaxation parameter is varying, or by appropriate modifications in the general case, leading to conditions on $(\gamma_n)_{n \in \mathbb{N}}$. For example, in the case where $\lambda = 1$, $\gamma = \gamma_n$ and $M = \partial \Phi$ (where $\Phi$ is proper convex and lower semicontinuous) converges if $\sum_n \gamma_n = +\infty$ (see, e.g., Thm. 28.1 Bauschke and Combettes (2017)).

The proximal point method was introduced (at least in the context of VIs) by Martinet (1970) who used it as an approximation algorithm. The proximal point method is globally convergent. However, it requires knowledge of the explicit form of the resolvent operator, which in many cases is not feasible.

**The Proximal Point Method for Variational Inequalities**

In the case of the VI,

$$M = F + N_C,$$

where $N_C$ is the normal cone of $C$. Then, the proximal point algorithm (3.25) (for $\gamma = \gamma_n$) becomes (upon multiplying by $\gamma_n$ which of course does not change the inclusion)

$$
\begin{aligned}
0 \in \gamma_n(F(x_{n+1}) + N_C(x_{n+1})) + (x_{n+1} - x_n) \iff \\
-\Big((x_{n+1} - x_n) + \gamma_n F(x_{n+1})\Big) \in \gamma_n N_C(x_{n+1}) = N_C(x_{n+1}) \\
\iff \Big(x_n - \gamma_n F(x_{n+1})\Big) - x_{n+1} \in N_C(x_{n+1}) \iff \\
x_{n+1} = \Pi_C(x_n - \gamma_n F(x_{n+1})).
\end{aligned}
\tag{3.27}
$$

Hence, for the special case of variational inequalities the proximal point method reduces to

$$
x_{n+1} = \Pi_C(x_n - \gamma_n F(x_{n+1})). \tag{3.28}
$$

This coincides with the proximal point algorithm (3.3) that was introduced in Sect. 3.2.

**Algorithm 3.20 (Proximal Point (Backward Projection) Algorithm)**

- Input $x_0 \in C$, and $(\gamma_n)_{n\in\mathbb{N}} \subset \mathbb{R}$, $\gamma_n > 0$, such that $\sum_{n\in\mathbb{N}} \gamma_n = +\infty$.
- Until a convergence criterion is met iterate

$$
x_{n+1} = \Pi_C(x_n - \gamma_n F(x_{n+1})).
$$

- When iteration stops return $x_n$.

**Remark 3.21 (The Inexact Proximal Method)** Rockafellar (1976) proposed an inexact variant of the proximal point method as $x_n + e_{n+1} \in x_{n+1} + \gamma_n M(x_{n+1})$, where $(e_n)_{n\in\mathbb{N}}$ is an error sequence, and has shown that if the error sequence is chosen so that $\sum_{n=0}^{\infty} \|e_n\| < \infty$, then $x_n \to x$ where $0 \in M(x)$. ⊕

### *3.3.3 Proximal Schemes Using Operator Splitting*

Since the resolvent operator $R_{\gamma M}$ is not always easily calculated, further approximation is required in order to render the use of the proximal scheme more practical. In many cases of interest, the operator $M$ can be split as the sum of two operators $M = A + B$, one of which may be easier to handle, for example, $A$ may be single-valued. Cases like this are usual in the study of VIs, where, for example, $M = F + N_C$, with $F$ a single-valued monotone operator. The basic philosophy behind operator splitting methods is to use an iterative scheme in which we first work with the operator $A$ and then with the operator $B$, whose resolvents may be easier to calculate, and iterate until convergence is reached.

In all cases below the problem to be considered is to solve the operator equation

$$0 \in (A + B)(x), \tag{3.29}$$

where $A$ and $B$ are maximal monotone operators.

#### 3.3.3.1 Forward-Backward Splitting (FBS)

The FBS algorithm was first proposed in the seminal publication Lions and Mercier (1979), which initiated the discussion of the use of splitting methods for finding the zeros of monotone operators (see also Bauschke and Combettes (2017)).

For the Forward-Backward splitting (FBS) scheme, assume that $A$ is single-valued, and express $A + B = \gamma^{-1}((I + \gamma B) - (I - \gamma A))$, $\gamma > 0$, and

$$\begin{aligned} 0 \in (A + B)(x) \iff & 0 \in \Big((I + \gamma B) - (I - \gamma A)\Big)(x) \\ \iff & \begin{cases} w \in (I + \gamma B)(x) \\ w = (I - \gamma A)(x). \end{cases} \end{aligned} \tag{3.30}$$

Recalling the definition of $R_{\gamma B}$, we see that the first inclusion of (3.30) implies that

$$x = (I + \gamma B)^{-1}(w) = R_{\gamma B}(w), \tag{3.31}$$

while the second equation of (3.30) yields

$$x = (I + \gamma B)^{-1}(I - \gamma A)(x) = R_{\gamma B}\Big((I - \gamma A)(x)\Big). \tag{3.32}$$

An alternative way is to see the RHS of above as a suggestion for updating $x$, which can be utilized in a relaxation scheme.

**Do Remember 3.22 (Splitting Schemes)**

*$A : X \to X$, $B : X \to 2^X$ maximal monotone, $\gamma > 0$, $\lambda_n > 0$.*
***Forward-Backward Splitting (FBS):***

$$x_{n+1} = (I + \gamma B)^{-1}(I - \gamma A)(x_n) = R_{\gamma B}\Big((I - \gamma A)(x_n)\Big). \tag{3.33}$$

***Relaxed Forward-Backward Splitting (RFBS):***

$$\begin{aligned} z_n &= x_n - \gamma A(x_n), \\ x_{n+1} &= x_n + \lambda_n(R_{\gamma B}(z_n) - x_n). \end{aligned} \tag{3.34}$$

For the convergence of the FBS scheme (see, e.g., Thm 26.14 Bauschke and Combettes (2017)), we have the following:

**Proposition 3.23** *Assume that $B : X \to 2^X$ is maximal monotone and that $A : X \to X$ is such that $\alpha A$ is $1/2$-averaged (or firmly non-expansive) for some $\alpha > 0$.*[4] *If $\sum_{n \in \mathbb{N}} \lambda_n(\delta - \lambda_n) = +\infty$, where $\lambda_n \in [0, \delta]$, $\delta = 2 - \frac{\gamma}{2\alpha}$, then the FBS relaxation scheme* (3.34) *converges to a point $x_o \in Zer(A + B)$, if $Zer(A + B) \neq \emptyset$. For the simpler FBS scheme* (3.33) *($\lambda_n = 1$) convergence requires $0 < \gamma \leq 2\alpha$.*

***Proof*** We start by noting that the fact that $\alpha A$ is $\frac{1}{2}$-averaged implies that $I - \gamma A$ is $\frac{\gamma}{2\alpha}$-averaged. Indeed, by the properties of $A$, there exists $N$ non-expansive such that $\alpha A = \frac{1}{2}I + \frac{1}{2}N$, so that $I - \gamma A = (1 - \frac{\gamma}{2\alpha})I + \frac{\gamma}{2\alpha}(-N)$, with $-N$ non-expansive.

By Proposition 3.17, $R_{\gamma B}$ is $\frac{1}{2}$-averaged, and hence, by Proposition 3.9 the composition $R_{\gamma B}(I - \gamma A)$ is $\theta$-averaged with $\theta = \frac{2\alpha}{4\alpha - \gamma}$; hence the FBS scheme with constant step reduces to the standard Krasnoselskii-Mann scheme and has the required convergence properties by Proposition 3.11.

The convergence of the relaxed scheme (3.34) follows by the relaxed Krasnoselskii-Mann scheme in Proposition 3.12. The constant step scheme (3.33) is the scheme (3.34) for the choice $\lambda_n = 1$, and the criterion for the choice of $\gamma$ follows from $\delta - 1 \geq 0$. □

**The Forward-Backward Splitting (FBS) Method for Variational Inequalities**
For the specific case of VIs considered here, the operator $M = F + N_C$, so using the above splitting and recalling that $R_{\gamma_k N_C} = \Pi_C$, we conclude that the Forward-Backward Splitting fixed point scheme for the VI becomes

$$x_{n+1} = \Pi_C(x_n - \gamma F(x_n)),$$

[4] Operators of this type are often called $\alpha$-co coercive operators and are such that $A^{-1}$ is $\alpha$-strongly monotone.

which is the standard projection scheme, presented in Algorithm 3.1. The convergence follows by Proposition 3.23.

#### 3.3.3.2 Backward-Forward Splitting (BFS)

The Backward-Forward Splitting scheme starts from the last equivalence in (3.30), now interpreted as

$$\begin{cases} x = R_{\gamma B}(w) \\ w = (I - \gamma A)(x) \end{cases} \iff \begin{cases} x = R_{\gamma B}(w) \\ w = (I - \gamma A)(R_{\gamma B}(w)). \end{cases} \tag{3.35}$$

**Do Remember 3.24 (Splitting Schemes—Continued from 3.22)**

***Backward-Forward Splitting:*** *For* $\gamma > 0$,

$$\begin{aligned} x_{n+1} &= R_{\gamma B}(w_n), \\ w_{n+1} &= (I - \gamma A)(x_{n+1}). \end{aligned} \tag{3.36}$$

**Backward-Forward Splitting for VIs**
For the specific case of VIs considered here, the operator $M = F + N_C$, so using the above splitting and recalling that $R_{\gamma_n N_C} = \Pi_C$, we conclude that the Backward-Forward Splitting fixed point scheme for the VI becomes

$$\begin{aligned} x_{n+1} &= \Pi_C(w_n), \\ w_{n+1} &= x_{n+1} - \gamma F(x_{n+1})). \end{aligned}$$

#### 3.3.3.3 The Douglas-Rachford Scheme

The starting point for the Douglas-Rachford scheme is the splitting

$$0 \in (A + B)(x) \iff 0 \in (I + \gamma A)(x) - (I - \gamma B)(x), \ \ \gamma > 0. \tag{3.37}$$

This scheme makes use of the resolvent and the Cayley (or reflection) operator (see Definition 3.13) and the identity (3.23), which allow us to express the splitting (3.37) as

$$0 \in (A + B)(x) \iff 0 \in (I + \gamma A)(x) - C_{\gamma B}(I + \gamma B)(x), \ \ \gamma > 0. \tag{3.38}$$

Since $R_{\gamma B}$ is single-valued, we choose an element

$$z \in (I + \gamma B)(x) \iff x = R_{\gamma B}(z) \tag{3.39}$$

and rewrite (3.38) as

$$\begin{aligned} 0 \in (A+B)(x) &\iff \begin{cases} z \in (I + \gamma B)(x), \\ w = C_{\gamma B}(z), \\ w \in (I + \gamma A)(x) \end{cases} \\ &\iff \begin{cases} x = R_{\gamma B}(z), \\ w = C_{\gamma B}(z), \\ x = R_{\gamma A}(w) \end{cases} \iff \begin{cases} x = R_{\gamma B}(z), \\ R_{\gamma B}(z) = R_{\gamma A}(C_{\gamma B}(z)). \end{cases} \end{aligned} \tag{3.40}$$

We now note that the last equation in the last equivalence of (3.40) can be expressed as a fixed point equation, in terms of $z$. By expressing the resolvents in terms of the Cayley reflection operators, we find that

$$\begin{aligned} R_{\gamma B}(z) = R_{\gamma A}(C_{\gamma B}(z)) &\iff \frac{1}{2}(I + C_{\gamma B})(z) = \frac{1}{2}(I + C_{\gamma A})\Big(C_{\gamma B}(z)\Big) \\ &\iff z = C_{\gamma A}(C_{\gamma B}(z)). \end{aligned}$$

Hence, we obtain the Douglas-Rachford splitting in terms of

$$\begin{aligned} 0 \in (A+B)(x) &\iff \begin{cases} x = R_{\gamma B}(z), \\ z = \Big(C_{\gamma A} C_{\gamma B}\Big)(z) \end{cases} \\ &\iff \begin{cases} x = R_{\gamma B}(z), \\ z = \Big(\frac{1}{2} I + \frac{1}{2} C_{\gamma A} C_{\gamma B}\Big)(z), \end{cases} \end{aligned} \tag{3.41}$$

for $\gamma > 0$, with the last step made so that the operator involved in the fixed point equation is a $1/2$-averaged operator (since $R_{\gamma A} R_{\gamma B}$ is not expansive). Therefore, the scheme consists of finding a fixed point $z$ of the operator $T := \frac{1}{2} I + \frac{1}{2} C_{\gamma A} C_{\gamma B}$ and then using it to obtain the solution $x$ in terms of $x = R_{\gamma B}(z)$.

The fixed point for $T$ is approximated by the Krasosnelskii-Mann iterative scheme (see (3.12)) or its relaxed version (see (3.18)). For example, scheme (3.12) yields

$$z_{n+1} = \left(\frac{1}{2}I + \frac{1}{2}C_{\gamma A}C_{\gamma B}\right)z_n \Longrightarrow \begin{cases} x_n = R_{\gamma B}(z_n), \\ z_{n+1} = z_n + \rho(R_{\gamma A}(2x_n - z_n) - x_n), \end{cases} \tag{3.42}$$

with $\rho = 1$, whereas scheme (3.18) will give a relaxed version of (3.42) with $\rho = \lambda_n$.

An equivalent way of expressing (3.42) is by noting that the Douglas-Rachford splitting (3.40) is in fact a compatibility condition for two different expressions for $x$, i.e., $x = R_{\gamma B}(z)$ and $x = R_{\gamma A}(w)$, with $w = C_{\gamma B}(z)$. We may then propose two different iterative approximation schemes for $x$, using each of the above expressions, monitor their difference, and iterate until the difference becomes negligible, with a relaxation step.

**Do Remember 3.25 (Douglas-Rachford (DRS) Scheme)**
*A, B maximal monotone (possibly multi-valued),* $\gamma > 0$, $\lambda_n > 0$.

$$\begin{aligned} w_{n+1} &= R_{\gamma B}(z_n), \\ x_{n+1} &= R_{\gamma A}(2w_{n+1} - z_n), \\ z_{n+1} &= z_n + \lambda_n(x_{n+1} - w_{n+1}). \end{aligned}$$

Variable step or other variants are possible. The following proposition (see, e.g., Cor. 28.1 Bauschke and Combettes (2017)) provides a convergence result.

**Proposition 3.26** *The DRS scheme converges to some* $x^* \in Zer(A + B)$ *(if this set is nonempty) if* $(\lambda_n)_{n\in\mathbb{N}} \subset (0, 2)$ *and* $\sum_{n\in\mathbb{N}} \lambda_n(2 - \lambda_n) = +\infty$.

***Proof*** The operators $C_{\gamma A}$ and $C_{\gamma B}$ are non-expansive (see Proposition 3.17(i)); hence the operator $C_{\gamma A}C_{\gamma B}$ is non-expansive, and the operator $\frac{1}{2}I + \frac{1}{2}C_{\gamma A}C_{\gamma B}$ is $\frac{1}{2}$-averaged.

We therefore see that the DRS scheme is in the general form of the variable step relaxation scheme treated in Proposition 3.12, and the convergence follows from that. □

**Remark 3.27 (The Peaceman-Ratchford Splitting (PRS))** The DRS scheme for the choice $\lambda_n = 2$ is called the Peaceman-Ratchford Splitting (PRS) scheme (see Prop. 26.13, Bauschke and Combettes (2017)). ⊕

**The Douglas-Rachford Splitting (DRS) Method for Variational Inequalities**
For the case of VIs where $M = A + B := F + N_C$, this scheme becomes:

**Algorithm 3.28 (Douglas-Rachford Scheme for VIs)**

- For $\gamma > 0$, $\lambda_n > 0$, until a stopping criterion is met iterate

$$\begin{aligned} w_{n+1} &= \Pi_C(z_n), \\ x_{n+1} &= R_{\gamma F}(2w_{n+1} - z_n), \\ z_{n+1} &= z_n + \lambda_n(x_{n+1} - w_{n+1}). \end{aligned}$$

- Return $x$.

Variable step size schemes or other variants are possible. The calculation of $R_{\gamma F}$ is the difficult part. In the case where $F = \partial\phi$ (the subdifferential or the gradient of a convex function), the $R_{\gamma F}$ is calculated in terms of the Moreau approximation, i.e., as

$$x_{n+1} = \arg\min_z \left( \phi(z) + \frac{1}{2\gamma}\|x - z\|^2 \right), \quad x = 2w_{n+1} - z_n.$$

#### 3.3.3.4 The Modified Forward-Backward Scheme

This scheme was first proposed by Tseng (2000) as a general scheme for locating the zeros of sums of monotone operators and has since gained popularity as a scheme for the solutions of variational inequalities.

**Do Remember 3.29 (Modified Forward-Backward Scheme (MFBS))**
*Under the assumption that both $A$ and $B$ are maximal monotone, with $A$ single-valued and $B$ possibly multi-valued, and such that $dom(B) \subset dom(A)$, MFBS is an alternative splitting leading to an iterative scheme of the form*

$$\begin{aligned} z_n &= (I + \gamma B)^{-1}(I - \gamma A)(x_n), \\ x_{n+1} &= \Pi_S(z_n - \gamma(A(z_n) - A(x_n)), \end{aligned} \tag{3.43}$$

*where $\gamma > 0$ and $S$ is an appropriate subset of $X$, chosen such that $X \subset dom(A)$ and $S \cap (A + B)^{-1}(0) \neq \emptyset$.*

Other variants are possible. Note also that $(I+\gamma B)^{-1} = R_{\gamma B}$. There is a certain freedom on the choice of the set $S$ (with the choice $S = X$ also allowed). A general rule is to choose $S$ so that:

1. Either $A$ is Lipschitz on $S \cup dom(B)$
2. Or $A$ is continuous on $dom(B)$ and $S \subset dom(B)$

Hence, if $A$ is Lipschitz on a closed convex set $K \subset dom(A)$ and $dom(B) \subset K$, then we may choose $S = K$. If $A$ is Lipschitz on the whole of the vector space $X$, we can choose $S = X$ and remove $\Pi_S$ from (3.43).

On the other hand, if $dom(B)$ is closed, then we may choose $S = dom(B)$. This example is particularly useful for variational inequalities, where $B = N_C$, so that $dom(B) = C$; hence we can choose $S = C$ in (3.43). This is also the case when $B = M + N_C$, where $M$ is a (possibly multi-valued) maximal monotone operator such that $C \subset dom(M)$.

For the convergence of the Tseng algorithm, it is not straightforward to apply directly the results on averaged operators on $(I - \gamma A)R_{\gamma B}(I - \gamma A) + \gamma A$, so instead we follow a direct approach to the convergence result, which is instructive in its own right. We also introduce a scheme for selecting the steps, which is often used in practice, called the generalized Armijo-Goldstein rule.

**Definition 3.30 (Generalized Armijo-Goldstein Step Size Rule)** The step sizes $\gamma$ (constant or not) are selected so that, for some $\theta \in (0, 1)$, fixed, at each iteration,

$$\gamma \|A(z) - A(x)\| \leq \theta \|z - x\|. \tag{3.44}$$

**Proposition 3.31** *Under the standing assumptions of this section and the step size rule* (3.44) *of Definition 3.30, the Tseng algorithm converges to some $x \in Zer(A + B)$ (strong if $dim(X) < \infty$, weak otherwise).*

***Proof*** The proof proceeds in several steps.
0. We collect some preliminary facts. Clearly, if we set

$$z = J(x, \gamma) := (I + \gamma B)^{-1}(I - \gamma A)(x), \quad \text{then} \quad \frac{1}{\gamma}(x - z) \in A(x) + B(z). \tag{3.45}$$

To see this recall that

$$\begin{aligned} z = (I + \gamma B)^{-1}(I - \gamma A)(x) &\implies \exists v \in B(z) \text{ such that } z + \gamma v = x - \gamma A(x) \\ &\implies \gamma^{-1}(x - z) = A(x) + v \in A(x) + B(z). \end{aligned} \tag{3.46}$$

By the monotonicity of $B$ for any $u \in A(x) + B(z)$, we have that

$$\langle \underbrace{\gamma^{-1}(x - z) - A(x)}_{\in B(z) \quad (3.45)} - \underbrace{(u - A(x))}_{\in B(x)}, z - x \rangle \geq 0,$$

which rearranged yields

$$\gamma^{-1}\|x - z\|^2 \leq \langle u, x - z \rangle \leq \|u\| \|x - z\| \implies \gamma^{-1}\|x - z\| \leq \|u\|$$
$$\leq \sup_{u \in A(x)+B(x)} \|u\|,$$

from which follows the fact that

$$x - J(x, \gamma) \to 0, \quad \text{as } \gamma \to 0. \tag{3.47}$$

1. Well-posedness of Armijo-Goldstein step size rule: Observation (3.47) is important as it shows the existence of a step $\gamma$ which is small enough such that the Armijo-Goldstein-like step size rule (3.44) holds, for any $x_n \notin S \cap Zer(A + B)$. Indeed, suppose that (3.44) could not be satisfied for any small $\gamma$. Then,

$$\gamma \|A(z_n) - A(x_n)\| \geq \theta \|z_n - x_n\|, \tag{3.48}$$

for all $\gamma > 0$ small, whereas, on the other hand,

$$\gamma^{-1}(x_n - z_n) \in A(x_n) + B(z_n). \tag{3.49}$$

Dividing (3.48) by $\gamma$ and passing to the limit $\gamma \to 0^+$, taking into account the fact that $x_n - z_n \to 0$ so that $\|A(x_n) - A(z_n)\| \to 0$ by continuity of $A$, in this limit, combined with (3.49) leads to the observation that $0 \in A(x_n) + B(x_n)$, which is a contradiction.
2. Let $w := z - \gamma(A(z) - A(x))$, and consider any $x^* \in S \cap Zer(A + B)$. Then, there exists an $\eta \geq 0$ such that

$$\|\Pi_S(w) - x^*\|^2 \leq \|w - x^*\|^2 = \|x - x^*\|^2 + \gamma^2\|A(z) - A(x)\|^2 - \|z - x\|^2 - 2\gamma\eta. \tag{3.50}$$

We prove this claim. The first inequality is easy from the non-expansive property of the projection operator. We now use the identity

$$\begin{aligned} \|x - x^*\|^2 &= \|x - z + z - w + w - x^*\|^2 \\ &= \|x - z\|^2 - \|z - w\|^2 + \|w - x^*\|^2 - 2\langle x - w, z - x^* \rangle, \end{aligned} \tag{3.51}$$

which can be easily verified by direct expansion of the LHS and the RHS. This is true for any $w \in X$; choosing $w = z - \gamma(A(z) - A(x))$ (as in the beginning of the step), we see that by (3.46)

$$\begin{aligned} x - w &= x - z + \gamma(A(z) - A(z)) = \gamma v + \gamma A(x) + \gamma(A(z) - A(x)) \\ &= \gamma v + \gamma A(z), \quad v \in B(z), \end{aligned} \tag{3.52}$$

whereas

$$x^* \in Zer(A+B) \implies \exists\, u \in B(x^*) \text{ such that } u + A(x^*) = 0. \tag{3.53}$$

Combining (3.52) and (3.53), we see that

$$\begin{aligned} 2\langle x - w, z - x^* \rangle &= 2\gamma \langle A(z) + v, z - x^* \rangle = 2\gamma \langle A(z) + v - \underbrace{(A(x^*) + u)}_{=0}, z - x^* \rangle \\ &=: 2\gamma\eta, \end{aligned} \tag{3.54}$$

with $\eta \geq 0$ by the monotonicity of $A$ and $B$ (note that $v \in B(z)$ by (3.46) and $u \in B(x^*)$ by (3.53)). Substituting(3.54) in (3.51) and rearranging yield (3.50).
3. Apply (3.50) to $w = w_n := z_n - \gamma(A(z_n) - A(x_n))$ and $x = x_n$, keeping in mind that $\Pi_S(w_n) = x_{n+1}$. This yields (omitting the last term)

$$\|x_{n+1} - x^*\|^2 \leq \|x_n - x^*\|^2 + \gamma^2 \|A(z_n) - A(x_n)\|^2 - \|z_n - x_n\|^2.$$

We can control the second term on the RHS of the above using the Armijo type rule (3.44) for the choice of step $\lambda$, which yields

$$\|x_{n+1} - x^*\|^2 \leq \|x_n - x^*\|^2 - (1 - \theta^2)\|z_n - x_n\|^2, \tag{3.55}$$

which for $\theta < 1$ leads to the Fejér property of $(x_n)_{n\in\mathbb{N}}$ with respect to $Zer(A+B)$,

$$\|x_{n+1} - x^*\| \leq \|x_n - x^*\|. \tag{3.56}$$

4. The convergence of the sequence $(x_n)_{n\in\mathbb{N}}$ arises from its Fejer property (3.56) by standard arguments, similar to those used in the proof of Proposition 3.11. For instance the Fejer property guarantees the boundedness of $(x_n)_{n\in\mathbb{N}}$, which by compactness arguments leads to (up to subsequences) the existence of a converging subsequence (strong if $X$ is finite dimensional and weak if $X$ is infinite dimensional) to some limit $x \in X$. Moreover, solving (3.55) with respect to $\|z_n - x_n\|^2$, adding over all $n = 1, \cdots, N$, and using the boundedness of $\|x_N - x^*\|^2$ yield that $\|z_n - x_n\| \to 0$ as $n \to \infty$. On the other hand, (3.49), for $x = x_n$, $z = z_n$ (and $\gamma = \gamma_n$, if we choose a variable step scheme), after moving the term $A(x_n)$ to the

LHS and adding the term $A(z_n)$ on both sides, leads to the fact that

$$\gamma^{-1}(x_n - z_n) + (A(z_n) - A(x_n)) \in (A + B)(z_n)$$
$$\Longrightarrow (z_n, \gamma^{-1}(x_n - z_n) + (A(z_n) - A(x_n))) \in Gr(A + B),$$

and passing to the convergent subsequences and using the observations made towards the end of step 1, along with the closedness of $Gr(A + B)$ (with a bit of care over technicalities required for the infinite dimensional case) leads to the result that $(x, 0) \in Gr(A + B)$, hence $0 \in (A + B)(x)$. Since the limit $x \in Zer(A + B)$, using again the Fejér property of $(x_n)_{n\in\mathbb{N}}$, using now $x^* = x$, we can show that the whole sequence $(x_n)_{n\in\mathbb{N}}$, and not just the extracted subsequence, converges to some $x \in Zer(A + B)$. □

**The Modified Forward-Backward Scheme for Variational Inequalities**

The modified forward-backward scheme of Tseng is directly applicable to VIs, by applying it to the operator $A + B = F + N_C$, in which case $R_{\gamma B} = (I + \gamma B)^{-1} = \Pi_C$. In this case the method reduces to the following:

**Algorithm 3.32 (MFBS for VIs)**

- For $\gamma > 0$, until a stopping criterion is met iterate

$$\begin{aligned} z_{n+1} &= \Pi_C(x_n - \gamma F(x_n)), \\ x_{n+1} &= \Pi_S(x_n - \gamma(F(z_n) - F(x_n))). \end{aligned} \tag{3.57}$$

- Return $x$.

The step size $\gamma$ can be variable, i.e., $\gamma = \gamma_n$. Note the small difference of (3.57) with the extragradient scheme (3.4).

## 3.4 Duality-Based Schemes

Duality based schemes stem from the idea of expressing the variational inequality first as an optimization problem and then as a saddle point problem using an appropriate duality formulation. Then, an approximation scheme for determining the saddle point is used to recover the solution of the original VI. That said, it is important to mention that there is not a unique duality scheme that can be used, and various notions of duality can be employed, depending on the nature of the problem.

### 3.4.1 Duality for Optimization Problems

Before moving to VIs, we recall some facts from the theory of duality for optimization problems. We refer to, e.g., Ekeland and Temam (1999) and Kravvaritis and Yannacopoulos (2020) for more details.

The general scheme for duality is to start from an optimization problem

$$\inf_{x \in C_X} f(x), \tag{3.58}$$

where $C_X$ is a closed convex subset of a vector space $X$ (either finite or infinite dimensional) called hereafter the primal problem (P), and try to express the restriction of the cost function $f$ on $C$ in terms of a new function $L : X \times Y \to \mathbb{R}$ (where $Y$ is another vector space), chosen so that

$$f(x) = \sup_{y \in C_Y} L(x, y), \tag{3.59}$$

where $C_Y \subset Y$ is a suitably chosen closed convex subset of $Y$ (possibly $C_Y = Y$). The function $L$ is considered as a perturbation of $f$ and is often called the Lagrangian.

Having obtained such a perturbation function $L$, the primal problem is then expressed as

$$(P) \quad \inf_{x \in C_X} f(x) = \inf_{x \in C_X} \sup_{y \in C_Y} L(x, y). \tag{3.60}$$

The next step is to invoke a general minimax theorem (such as, e.g., Sion's minimax theorem Sion (1958)) and invert the order of the inf and the sup operation in (3.60). This can be done under certain conditions[5] on the Lagrangian $L$ and on the sets $C_X$ and $C_Y$. Working formally, we can define the so-called dual problem:

$$(D) \quad \sup_{y \in C_Y} \inf_{x \in C_X} L(x, y) =: \sup_{y \in C_Y} D(y), \tag{3.61}$$

where $D : Y \to \mathbb{R}$ is the dual function defined by

$$D(y) := \inf_{x \in C_X} L(x, y).$$

Often, the choice of Lagrangian is made so that the resulting optimizations in the primal and dual problem are on more convenient sets than $C_X$ and $C_Y$. This point will be clarified in the forthcoming examples.

[5] Essentially convexity and concavity of $L$ as a function of $x$ and $y$, respectively, and some kind of compactness (strong or weak of some sort) for $C_X, C_Y$.

As already mentioned, unless a minimax theorem is applicable for $L$, $C_X$, and $C_Y$, the value for the primal ($V_P$) and the dual problem ($V_D$) is not equal. In general, the so-called weak duality holds

$$V_D \leq V_P. \tag{3.62}$$

This follows easily by carefully comparing the values of the inf and sup operations. This can be interesting in its own right, as it can be used to provide lower bounds for the primal problem, as long as the dual problem is easy to handle. It is worth mentioning here that often the dual problem is better behaved than the primal, so that (3.62) can indeed offer useful a priori bounds.

If, moreover, the minimax theorem is applicable, then we obtain the strong duality result

$$V_D = V_P \iff \underbrace{\inf_{x \in C_X} f(x) = \inf_{x \in C_X} \sup_{y \in C_Y} L(x, y)}_{(P)} = \underbrace{\sup_{y \in C_Y} \inf_{x \in C_X} L(x, y)}_{(D)}. \tag{3.63}$$

In this case we obtain a useful representation of the solution of the primal problem $x^*$ in terms of the solution of the dual problem $y^*$

$$y^* \in \arg\max_{y \in C_Y} \inf_{x \in C_X} L(x, y) = \arg\max_{y \in C_Y} D(y).$$

This representation arises from the observation that $(x^*, y^*)$ is a saddle point of the Lagrangian function $L$, i.e.,

$$L(x^*, y) \leq L(x^*, y^*) \leq L(x, y^*), \quad \forall (x, y) \in C_X \times C_Y. \tag{3.64}$$

That allows us to characterize the point $(x^*, y^*)$ in terms of the fixed point

$$\begin{aligned} x^* &\in \arg\min_{x \in C_X} L(x, y^*), \\ y^* &\in \arg\max_{y \in C_Y} L(x^*, y) \end{aligned} \tag{3.65}$$

or in terms of the subdifferential operator as the first order condition

$$\begin{aligned} 0 &\in \partial_x L(x^*, y^*), \\ 0 &\in \partial_y(-L(x^*, y^*)), \end{aligned} \tag{3.66}$$

where $\partial_x L(x, y)$ is the subdifferential of the function $x \mapsto L(x, y)$ for $y$ fixed (similarly for $\partial_y$) and the minus sign in the second relation is needed as we consider a maximization problem and we consider $y \mapsto L(x, y)$ as a concave function.

The characterizations (3.65) or (3.66) lead to many useful numerical schemes, called duality based schemes. These schemes are particularly useful, especially for choices of the Lagrangian function for which the minimization over $x \in C_X$ can be replaced with minimization over $X$, and the set $C_Y$ has a nice structure. Operator splitting schemes such as those introduced in Sect. 3.3.3 can be used for both the primal and the dual part of formulation (3.66).

**Do Remember 3.33 (Duality Schemes)** *For a suitable $L : X \times Y \to \mathbb{R}$, express $f(x) = \sup_{y \in Y} L(x, y)$, so that*

$$V_P := \inf_{x \in X} f(x) = \inf_{x \in X} \sup_{y \in Y} L(x, y) \geq \sup_{y \in Y} \inf_{x \in X} L(x, y) =: V_D.$$

*Subject to conditions (e.g., if minimax applies), then $V_P = V_D$, and the solution $x^*$ of $(P)$ is characterized in terms of the pair $(x^*, y^*) \in X \times Y$, where $(x^*, y^*)$ is a saddle point of $L$.*

*Constraints can be introduced by the appropriate choice of $f$.*

**Example 3.34 (Lagrangian Duality)** Lagrangian duality is very popular for optimization problems of the form (3.58) with $C_X$ characterized in terms of $m$ inequality constraints,[6] as

$$C_X = \{x \in X \ : \ g_i(x) \leq 0, \ g_i : X \to \mathbb{R}, \ \text{convex}, \ i = 1, \cdots, m\}. \tag{3.67}$$

In this case we set $C_Y := \mathbb{R}^m_+$, and defining

$$\begin{aligned} L(x, y) &= f(x) + \sum_{i=1}^{m} y_i g_i(x) = f(x) + \langle y, g(x) \rangle_{\mathbb{R}^m}, \\ y &= (y_1, \cdots, y_m) \in C_Y := \mathbb{R}^m_+, \quad g(x) = (g_1(x), \cdots, g_m(x)), \end{aligned} \tag{3.68}$$

we easily see that

$$\sup_{y \in C_Y} L(x, y) = \begin{cases} f(x) \text{ if } x \in C_X, \\ +\infty \text{ if } x \notin C_X. \end{cases} \tag{3.69}$$

Hence,

$$\inf_{x \in C_X} f(x) = \inf_{x \in X} \sup_{y \in C_Y} L(x, y), \tag{3.70}$$

[6] $C_X$. follows by minimal assumptions on $g_i$.

leading under suitable conditions (the generalized Slater conditions) to the saddle point representation.

The dual problem becomes

$$\sup_{y \in C_Y} \inf_{x \in X} L(x, y) =: \sup_{y \in C_Y} D(y), \tag{3.71}$$

with the inner problem, leading to the definition of the dual function $D$ being greatly simplified, as it now is an optimization problem over the whole vector space $X$ rather than on the closed convex subset $C_X$. Moreover, while the dual problem is still a constrained problem, the constraints set is $C_Y = \mathbb{R}^m_+$ which is very convenient to handle, using, e.g., a projection approach, since the projection operator for this set is explicit. The above construction can be generalized if $X$ is infinite dimensional, but with $C_X$ as in (3.67). ◁

**Example 3.35 (Fenchel Duality)** Another popular duality approach is the so-called Fenchel duality which uses the concept of the Fenchel conjugate function. Consider the problem

$$\inf_{x \in X} \Big( f(x) + \Phi(x) \Big), \tag{3.72}$$

where $X$ is a Hilbert space and $\Phi : X \to \mathbb{R} \cup \{+\infty\}$ is a convex lower semicontinuous function. Moreover, we may choose $\Phi(x) = I_{C_X}(x)$, the convex indicator function of a closed convex set, hence turning (3.72) into a constrained optimization problem on $C_X$, or even consider (3.72) on $X \cap C_X$ for general $\Phi$.

In the above setting the Fenchel conjugate function of $\Phi$ is a function $\Phi^* : X' \to \mathbb{R} \cup \{+\infty\}$ defined by

$$\Phi^*(y) = \sup_{x \in X} \Big( \langle y, x \rangle - \Phi(x) \Big), \tag{3.73}$$

while the lower semicontinuity of $\Phi$ yields

$$\Phi(x) = \sup_{y \in X'} \Big( \langle y, x \rangle - \Phi^*(y) \Big). \tag{3.74}$$

In both (3.73) and (3.74), $\langle \cdot, \cdot \rangle$ is the duality pairing between $X'$ and $X$.[7] Moreover, the following hold:

$$\begin{aligned} \Phi(x) + \Phi^*(y) = \langle y, x \rangle, &\iff x \in \partial \Phi^*(y), \\ y \in \partial \Phi(x) &\iff x \in \partial \Phi^*(y). \end{aligned} \tag{3.75}$$

[7] If $dim(X) < \infty$, we can assume $X' \simeq X$, and the inversion is conceptually clear. In the infinite dimensional case some subtle functional analysis arguments are required (see, e.g., Kravvaritis and Yannacopoulos (2020), Kyritsi-Yiallourou and Papageorgiou (2009)).

With this information at hand, setting the space $Y$ in our general duality construction $Y = X'$, we can express the primal problem as

$$\begin{aligned}\inf_{x\in X}\Big(f(x)+\Phi(x)\Big) &= \inf_{x\in X}\Big(f(x)+\sup_{y\in Y=X'}\Big(\langle y,x\rangle-\Phi^*(y)\Big)\Big)\\ &=\inf_{x\in X}\sup_{y\in Y=X'}\Big(f(x)+\langle y,x\rangle-\Phi^*(y)\Big)=\inf_{x\in X}\sup_{y\in Y=X'}L(x,y),\end{aligned}\tag{3.76}$$

for the Lagrangian function

$$L(x,y)=f(x)+\langle y,x\rangle-\Phi^*(y).$$

Subject to appropriate conditions strong duality holds, with the dual problem being

$$\sup_{y\in Y=X}\inf_{x\in X}L(x,y)=\sup_{y\in Y=X'}\inf_{x\in X}\Big(f(x)+\langle y,x\rangle-\Phi^*(y)\Big)=:\sup_{y\in Y=X'}D(y),\tag{3.77}$$

with the dual function defined as

$$D(y):=\inf_{x\in X}\Big(f(x)+\langle y,x\rangle-\Phi^*(y)\Big).\tag{3.78}$$

Note that $D(y)$ can be expressed in terms of a Fenchel transform as well

$$D(y)=\inf_{x\in X}\Big(f(x)+\langle y,x\rangle\Big)-\Phi^*(y)=-\sup_{x\in X}\Big(\langle -y,x\rangle-f(x)\Big)-\Phi^*(y).$$

More general versions involving $\Phi(Bx)$, where $\Phi : Z \to \mathbb{R}\cup\{+\infty\}$, and $B : X \to Z$, a linear operator between suitable Hilbert spaces $X, Z$, are possible.[8] $\triangleleft$

### 3.4.2 Duality for Variational Inequalities

Duality results for variational inequalities can now follow easily by observing that certain types of variational inequalities can be rephrased in terms of optimization problems. Some important cases that fall in this category are the following:

(i) Problems of the form

$$\langle F(x), z-x\rangle \geq 0,\ \ \forall z\in C,\tag{3.79}$$

[8] In which case $Y = Z'$ and $L(x, y) = f(x) + \langle B^*y, x\rangle - \Phi^*(y)$, where $B^* : Z' \to X'$ the adjoint operator.

in the case where $F : \mathbb{R}^d \to \mathbb{R}^d$ can be expressed in terms of the gradient of a suitable function $\Phi : \mathbb{R}^d \to \mathbb{R}$, as $F = D\Phi$. Then, (3.79) is equivalent to the minimization problem

$$\min_{x \in C} \Phi(x), \tag{3.80}$$

which can then be treated using a suitable duality approach, depending on the nature of $C$.

(ii) The situation in (i) can be extended for the infinite dimensional case, for problems of the form (3.79) where $F : X \to Y$ with $X$ and $Y$ appropriate Hilbert (or Banach) function spaces, if $F = D\Phi$, the Gâteaux derivative of an appropriate functional $\Phi : X \to \mathbb{R}$, and $K \subset X$ is a closed convex set. Many of the problems in Sect. 2.6 can be considered as special cases of this type, for example, the case where $F(u) = -\Delta u + k^2 u$ and $K = \{u \in W^{1,2}(D) : u \geq \psi\}$ reduces to the minimization of the Dirichlet type functional $u \mapsto \Phi(u) := \int_D (|Du|^2 + k^2 u^2) dx$ .

(iii) Problems of the form

$$x \in C \; : \; \mathbf{a}(x, z - x) - \langle f, x \rangle + \Phi(z) - \Phi(x) \geq 0, \;\; \forall z \in C, \tag{3.81}$$

where $\mathbf{a} : X \times X \to \mathbb{R}$ is a symmetric bilinear continuous and coercive form, $\Phi$ is a proper lower semicontinuous convex function, and $X$ is either finite or infinite dimensional. Problem (3.81) is related to a quasi-variational inequality (QVI) and reduces to the standard VI for $\Phi = I_C$. Under the stated assumptions on $\mathbf{a}$, working as in Theorem 2.62(ii), problem (3.81) is equivalent to the minimization problem

$$\inf_{x \in X} f(x), \;\; \text{where} \;\; f(x) = \frac{1}{2} a(x, x) - \langle f, x \rangle + \Phi(x). \tag{3.82}$$

(iv) VIs can be transformed into saddle point problems, either in terms of the merit or gap function (see Sect. 1.1.2) or using monotonicity techniques such as Minty's reformulation (see Sect. 2.3.2, Proposition 2.27).

Any of the duality schemes presented in Sect. 3.4.1 can be used for the optimization problem corresponding to the VI at hand.

### *3.4.3 Numerical Optimization Problems for Saddle Point Problems*

Using the above duality schemes (either Lagrangian or Fenchel) and the corresponding saddle point formulation in combination with any of the numerical methods presented in Sect. 3.3, we can construct numerical schemes for saddle point problems (related to either VIs or more general ones), based on either formulation (3.65) or (3.66). These typically consist of a coupled descent/proximal type minimization scheme for the primal variable and an ascent/proximal type maximization scheme for the dual variable, with or without splitting.

**Do Remember 3.36 (Primal-Dual Schemes)**
***Primal descent—dual ascent scheme:*** *For* $\gamma > 0$,

$$\begin{cases} x_{n+1} \in x_n - \gamma \partial_x L(x_n, y_n), \\ y_{n+1} \in y_n - \gamma \partial_y (-L(x_n, y_n)). \end{cases} \tag{3.83}$$

***Saddle point proximal scheme:*** *For* $\gamma > 0$,

$$\begin{cases} x_{n+1} \in x_n - \gamma \partial_x L(x_{n+1}, y_n), \\ y_{n+1} \in y_n - \gamma \partial_y (-L(x_{n+1}, y_{n+1})) \end{cases} \iff \begin{cases} x_{n+1} = R_{\gamma \partial_x L}(x_n, y_n), \\ y_{n+1} = R_{\gamma \partial_y (-L)}(x_{n+1}, y_n). \end{cases} \tag{3.84}$$

For smooth $L$ (3.84) reduces to gradient descent for $x$ and gradient ascent for $y$. Other variants are of course possible, and a large literature is available. Here we choose to present a small sample, to provide the reader with the idea of implementation of the concepts of duality in the construction of concrete numerical schemes.

#### 3.4.3.1 The Alternating Direction Method of Multipliers (ADMMs)

This method (see Glowinski (2014)) is a popular method for minimization of separable convex optimization problems, subject to linear constraints, as well as the corresponding VIs. There are various versions and extensions of this method, e.g., including linear inequality constraints, etc. It uses the concept of augmented Lagrangian, hence falling into the realm of a class of methods called augmented Lagrangian methods (ALM), and also combines some operator splitting ideas (see, e.g., He et al. (2023) and the references therein).

A typical problem for which the ADMM method can be very useful is the problem of

$$\min_{x_1 \in X_1,\, x_2 \in X_2} (f_1(x_1) + f_2(x_2)) \text{ subject to } A_1 x_1 + A_2 x_2 = b, \tag{3.85}$$

which under conditions is directly related to a VI formulation (see Sect. 1.1.2). In (3.85), $x_i \in X_i$, $i = 1, 2$, are two sets of variables, each living on a (possibly) different Hilbert space $X_i$, $f_i : X_i \to \mathbb{R}$ are convex functions, and the solution is constrained by a linear constraint, defined in terms of the linear mappings $A_i$ :

$X_i \to Z$, where $Z$ is another Hilbert space and $b \in Z$. Without too much loss of generality, we can fix $X_i = \mathbb{R}^{d_i}$, $i = 1, 2$, and $Z = \mathbb{R}^m$, in which case $A_i$ are represented as matrices $A_i \in \mathbb{R}^{m \times d_i}$. Problems of this form are encountered in many applications, e.g., in the sharing of common resources, etc.

To obtain the ADMM method, we start with the augmented Lagrangian formulation of (3.85), which allows us to rewrite the original problem in terms of a saddle point formulation for the Lagrangian function

$$L(x_1, x_2; y) = f_1(x_1) + f_2(x_2) - \langle y, A_1 x_1 + A_2 x_2 - b \rangle_{\mathbb{R}^m}. \tag{3.86}$$

This maximization problem over $y \in \mathbb{R}^m$ is a linear problem, which is only well-posed if the linear constraint holds. The ADMM method is an iterative method for the corresponding saddle point problem of the form:

**Algorithm 3.37 (Alternating Direction Multipliers Method)**

- For $\gamma > 0, \beta > 0$, until a stopping criterion is met iterate

$$\begin{cases} x_{1,n+1} \in \arg\min_{x_1 \in X_1} L(x_1, x_{2,n}, y_n) + \underbrace{\frac{\beta}{2}\|A_1 x_1 + A_2 x_{2,n} - b\|^2}, \\ x_{2,n+1} \in \arg\min_{x_2 \in X_2} L(x_{1,n+1}, x_{2,}, y_n) + \underbrace{\frac{\beta}{2}\|A_1 x_{1,n+1} + A_2 x_2 - b\|^2}, \\ y_{n+1} \in \arg\max_{y \in Y} L(x_{1,n+1}, x_{2,n+1}, y) - \underbrace{\frac{1}{2\beta}\|y - y_n\|^2} \\ \quad = y_n - \beta(A_1 x_{1,n+1} + A_2 x_{2,n+1} - b). \end{cases} \tag{3.87}$$

- Return $(x_1, x_2)$.

Note that the dual part of (3.87) is in some sense an ascent scheme for the dual maximization problem. The extra (underbraced) terms are quadratic regularization terms, which provide a better behaviour for the linear dual part (the quadratic regularization term in $y$ leads to an inner solution to the linear optimization problem over $y$). In fact system (3.87) corresponds to a modified saddle point problem for a new regularized Lagrangian, called the augmented Lagrangian method (ALM). Clearly, in method (3.87) an operator splitting scheme has been applied (note that in each step we fix all the other variables but one and optimize over it).

#### 3.4.3.2 The Auslender and Teboulle (2000) Method

The Auslender and Teboulle (2000) method is a duality method based on the concept of merit functions (see Sect. 1.1.2) and Lagrangian duality (see Example 3.34).

Consider the variational inequality

$$\begin{aligned} &x \in C \ : \ \langle F(x), z - x \rangle \geq 0, \ \ \forall \, z \in C, \\ &C = \{x \in X \ : \ g_i(x) \leq 0, \ i = 1, \cdots, m\}, \ \ g_i : X \to \mathbb{R}, \ \ \text{convex}. \end{aligned} \tag{3.88}$$

Any solution of (3.88), $x^* \in C$, is also a solution of the optimization problem

$$\min_{z \in C} \Phi(z; x^*), \quad \text{where} \ \ \Phi(z; x) := \langle F(x^*), z - x^* \rangle, \tag{3.89}$$

with the objective function $z \mapsto \Phi(z; x)$ depending on the unknown $x^*$. Following Auslender and Teboulle (2000) (focusing on $X = \mathbb{R}^d$ and using similar steps as those in Sect. 1.1.2), we introduce the Lagrangian function for problem (3.89), $L : X \times \mathbb{R}^m_+ \to \mathbb{R}$, by

$$L(x, y; x^*) = \langle F(x^*), x - x^* \rangle + \sum_{i=1}^{m} y_i g_i(x) \tag{3.90}$$

(if $F$ is multi-valued, replace $F(x^*)$ by $z^* \in F(x^*)$, whereas if the $g_i$ are extended functions, we can work in $E = \bigcap_{i=1}^{m} dom(g_i)$). As in Example 3.34, we express the objective function of problem (3.89) in terms of

$$\sup_{y \in \mathbb{R}^m_+} L(x, y; x^*) = \langle F(x^*), x - x^* \rangle + I_C,$$

so that problem (3.89) can be written as

$$\inf_{x \in X} \sup_{y \in \mathbb{R}^m_+} L(x, y; x^*). \tag{3.91}$$

This leads to the dual problem

$$\sup_{y \in \mathbb{R}^m_+} \inf_{x \in X} L(x, y; x^*). \tag{3.92}$$

Moreover, if strong duality holds, the solution of the VI (3.88) can be identified in terms of the saddle point $(x^*, y^*) \in X \times \mathbb{R}^m_+$ of the Lagrangian function $L(x, y; x^*)$ with a corresponding formulation in terms of the Karush-Kuhn-Tucker

(KKT) conditions.[9] This can be identified in terms of the first order conditions, which for the particular Lagrangian are of the form

$$\begin{aligned} 0 &\in F(x^*) + \sum_{i=1}^{m} y_i^* \partial g_i(x^*), \\ 0 &\in -g(x^*) + N_{\mathbb{R}^m_+}(y^*), \end{aligned} \tag{3.93}$$

where by $N_{\mathbb{R}^m_+}(y^*)$ we denote (as usual) the normal cone of the set $\mathbb{R}^m_+$ and $\partial g_i(x^*)$ are the subdifferentials of the functions $g_i$ at the point $x^*$ (if the $g_i$ are smooth, then $\partial g_i(x^*) = \{Dg_i(x^*)\}$). Note that (3.93) is nothing else but the Karush-Kuhn-Tucker optimality conditions, written in an equivalent form, which reduces to finding the zero of a system of monotone operator equations. Clearly, the proximal methods introduced in Sect. 3.3.3 can be used in determining the solution to (3.93) and hence to the original problem (3.88).

A way to solve (3.93) is in terms of a proximal scheme for the variable $w = (x, y)$. For that we reformulate (3.93) in terms of the operator equation

$$0 \in T(w), \quad w = (x, y) \in X \times \mathbb{R}^m_+, \tag{3.94}$$

where $T : X \times \mathbb{R}^m_+ \to X \times \mathbb{R}^m_+$ with $T = (T_1, T_2)$ is the operator defined by

$$\begin{aligned} T_1 : X \times \mathbb{R}^m_+ \to X \quad &: \quad (x, y) \mapsto T_1(x, y) := F(x) + \sum_{i=1}^{m} y_i \partial g_i(x), \\ T_2 : X \times \mathbb{R}^m_+ \to \mathbb{R}^m_+ \quad &: \quad (x, y) \mapsto T_2(x, y) := -g(x) + N_{\mathbb{R}^m_+}(y). \end{aligned} \tag{3.95}$$

The operator $T$ is maximal monotone if $F$ is maximal monotone and such that $dom(F) \cap (\cap_{i=1}^m dom(g_i)) \neq \emptyset$ (see Prop. 3.2 in Auslender and Teboulle (2000)).[10]

A possible scheme for solving (3.94) could be the proximal scheme

$$\begin{aligned} w_{n+1} = R_{\gamma T}(w_n) \iff &\begin{cases} x_{n+1} = R_{\gamma T_1}(x_n, y_n), \\ y_{n+1} = R_{\gamma T_2}(x_n, y_n), \end{cases} \iff \\ &\begin{cases} 0 \in \frac{x_{n+1} - x_n}{\gamma} + F(x_{n+1}) + \langle y_{n+1}, \partial g(x_{n+1}) \rangle, \\ 0 \in \frac{y_{n+1} - y_n}{\gamma} - g(x_{n+1}) + N_{\mathbb{R}^m_+}(y_{n+1}), \end{cases} \end{aligned} \tag{3.96}$$

---

[9] The KKT formulation for this problem and $X = \mathbb{R}^d$ is related to the Slater condition of existence of an $x_o \in X$ such that $g_i(x_o) < 0$, $i = 1, \cdots, m$.

[10] Recall that $dom(f) := \{x \in X \ : f(x) < \infty\}$.

which upon using Proposition 2.16(ii) for the last inclusion leads to:

**Algorithm 3.38 (Primal-Dual Proximal Scheme I (Auslender-Tebulle))**

- For $\gamma > 0$, until a stopping criterion is met iterate

$$\begin{cases} 0 \in \frac{x_{n+1}-x_n}{\gamma} + F(x_{n+1}) + \langle y_{n+1}, \partial g(x_{n+1})\rangle, \\ \quad y_{n+1} = \Pi_{\mathbb{R}^m_+}(y_n + \gamma g(x_n)). \end{cases}$$

- Return $x$.

#### 3.4.3.3 The Bermúdez and Moreno (1981) Method

The method proposed by Bermúdez and Moreno (1981) is a popular duality method, based on Fenchel duality, for solving variational inequalities of the form

$$x \in X \ : \ \langle A(x), z - x\rangle - \langle f, z - x\rangle + \Psi(z) - \Psi(x) \geq 0, \ \ \forall z \in X, \tag{3.97}$$

where $X$ is a Hilbert space (either finite or infinite dimensional), $\Psi : X \to \mathbb{R} \cup \{+\infty\}$ is a proper lower semicontinuous convex function (i.e., not taking the value $+\infty$ everywhere), $A : X \to X'$ is a monotone operator, and $f \in X$ or $X'$. Note that this is a generalization of the standard form of variational inequality if we choose $\Psi = I_C$, the convex indicator function for the closed convex set $C$.[11] By $\langle \cdot, \cdot\rangle$ we denote either $\langle \cdot, \cdot\rangle_{X',X}$ or $\langle \cdot, \cdot\rangle_X$ (depending on the nature of the elements involved).

The starting point of duality based schemes is the observation that the VI (3.97) is equivalent to a minimization problem, on which a duality approach is applied. For the VI (3.97), the relevant minimization problem is

$$\min_{x\in X} \Phi(x), \quad \text{where } \Phi(x) := \frac{1}{2}\langle A(x), x\rangle - \langle f, x\rangle + \Psi(x), \ \forall x \in X. \tag{3.98}$$

For the Lagrangian formulation of (3.97), one may use the concept of the Fenchel duality. Working exactly as in Example 3.35, we can bring the problem in a saddle point formulation in terms of the Lagrangian function

$$L(x, y) := \frac{1}{2}\langle A(x), x\rangle - \langle f, x\rangle + \langle y, x\rangle - \Psi^*(y), \tag{3.99}$$

[11] In the original version of the problem, the constraint is of the form $\Psi(z_0 + Bz)$, where $\Psi : Z \to \mathbb{R} \cup \{+\infty\}$, $B : X \to Z$ bounded, and $Z$ an appropriate Hilbert space, but here we stick to its simpler version.

characterizing the solution $x^*$ in terms of the pair $(x^*, y^*)$ such that

$$\begin{cases} x^* \in \arg\min_{x\in X} L(x, y^*), \\ y^* \in \arg\max_{y\in X'} L(x^*, y), \end{cases} \implies \begin{cases} A(x^*) + y^* = f, \\ y^* \in \partial\Psi(x^*), \end{cases} \tag{3.100}$$

where by $\partial\Psi(x^*)$ we denote the subdifferential of the function $\Psi$.

One of the problems with the first order condition system (3.100), and the corresponding iterative scheme, is the fact that the subdifferential $M = \partial\Psi$ is a multi-valued operator, so in principle there is some ambiguity as to which element in the subdifferential to choose as the next iteration for $y$. One of the novelties of the Bermudez-Moreno scheme is the use of an approximation procedure for the subdifferential, the Moreau-Yosida approximation, $Y_{\gamma,M} := \gamma^{-1}(I - R_{\gamma M})$, $\gamma > 0$ (see Definition 3.13) that leads to a single-value approximate operator for the subdifferential at each iteration.

Using Example 3.16 (setting $x = x^* + \gamma y^*$, $y = y^*$), we express (3.100) in the equivalent form

$$\begin{aligned} A(x^*) + y^* &= f, \\ y^* &= Y_{\gamma,\partial\Psi}(x^* + \gamma y^*), \quad \gamma > 0. \end{aligned} \tag{3.101}$$

The Bermudez-Moreno algorithm can then be expressed as the fixed point iteration scheme.

**Algorithm 3.39 (Primal-Dual Proximal Scheme II (Bermudez-Moreno))**

- For $\gamma_n > 0$, until a stopping criterion is met, iterate

$$\begin{aligned} Ax_n + y_n &= f, \\ y_{n+1} &= Y_{\gamma_n,\partial\Psi}(x_n + \gamma_n y_n). \end{aligned} \tag{3.102}$$

- Return $x$.

## 3.5 Fully Worked Example: Numerics for Pricing American and Real Options

We now return to the problem of pricing American or real options introduced in Sect. 1.2.3 which is equivalent to a VI formulation according to Proposition 1.25, restricting for simplicity to the GMB model, working either with the operator (1.37) or (1.47) (after a suitable transformation). Now the corresponding mapping $F$, in the VI formulation, is a mapping $A$ between infinite dimensional spaces (see Sect. 2.5.1 onwards) which for numerical approximation has to be replaced by finite dimensional approximations. To further simplify the exposition, assume that $D \subset \mathbb{R}$, and use discretization, i.e., replace the function $u : D \to \mathbb{R}$ by the vector $x = (u(s_i), \ : i \in \{1, \cdots, d\}) \subset \mathbb{R}^d$, where $P := \{s_i, \ i = 1, \cdots, d\} \subset D$ is a partition of $D \subset \mathbb{R}$. Note that this is a well-defined procedure, since we have shown (see, e.g., Sect. 2.6.5) that the solution of the VI is a continuous function, so that the pointwise evaluation $u(s_i)$ at the points of the partition $P$ makes sense.

The numerical treatment of VIs requires a finite dimensional approximation of the action of the operator $A$ on the unknown function $u$, i.e., the function $Au$. This can be effected by approximating $Au$ as a function $F : \mathbb{R}^d \to \mathbb{R}^d$, acting on $x \in \mathbb{R}^d$, the finite dimensional approximation of the unknown function $u$, or equivalently to $x$, the finite dimensional approximation of $u$. There are various ways of doing this, perhaps the simpler would be to replace the derivatives $u'$, with finite dimensional approximations, using a similar discretization procedure, i.e., replace $u'$ by $z = (u'(s_1), \cdots, u'(s_d)) \subset \mathbb{R}^d$, where for each $i$ we approximate $u'(s_i) \approx \frac{1}{h_i}(u(s_{i+1}) - u(s_i)) = \frac{1}{h_i}(x_{i+1} - x_i)$, where $h_i = s_{i+1} - s_i$.[12] A similar approximation can be applied for the second derivatives, a common choice being

$$u''(s_i) \approx \frac{1}{h^2}(u(s_{i+1}) - 2u(s_i) + u(s_{i-1})) = \frac{1}{h^2}(x_{i+1} - 2x_i + x_{i-1}), \tag{3.103}$$

which can be justified by a Taylor expansion argument for the function $u$ (as long as it is sufficiently regular). Higher order approximations are also possible. Using the simple approximation (3.103) above, we can interpret the finite dimensional approximation of $Au$ as

$$\begin{aligned} (Au)(s_i) = -u''(s_i) + k^2 u(s_i) &\approx -\frac{1}{h^2}(u(s_{i+1}) - 2u(s_i) + u(s_{i-1})) + k^2 u(s_i) \\ &= -\frac{1}{h^2}(x_{i+1} - 2x_i + x_{i-1}) + k^2 x_i, \ \ i = 1, \cdots, d, \end{aligned} \tag{3.104}$$

which can be written in compact form as

$$F(x) = A_d x := (L + k^2 I)x, \ \ L = \frac{1}{h^2} \begin{pmatrix} 2 & -1 & 0 & 0 & \cdots & 0 \\ -1 & 2 & -1 & 0 & \cdots & 0 \\ 0 & -1 & 2 & -1 & \cdots & 0 \\ 0 & 0 & -1 & 2 & \cdots & 0 \\ & & & & \ddots & \vdots \\ 0 & 0 & 0 & 0 & \cdots & -1 \ 2 \end{pmatrix}, \ \ x = \begin{pmatrix} x_1 \\ \vdots \\ x_d \end{pmatrix}.$$

[12] Alternatively, we may use some centered approximation, e.g., for a uniform partition, $h_i = h$, $u'(s_i) \approx \frac{1}{2h}(u_{i+1} - u_{i-1})$.

The matrix $L$ is a tridiagonal $d \times d$ matrix, whose rows, except for the first and last, represent the action of the operator of differentiating the function $u$ twice, on the vector $x$ containing the values of $u$ on the chosen partition. Note that all these rows are identical, just shifting the block $(-1, 2, 1)$ one position to the right as we move from one row to the next. The first and last rows are different. Recall that the VI is complemented with boundary conditions. Assuming without loss of generality that the boundary conditions are homogeneous Dirichlet, we know the value of the unknown function on the boundary $u(0) = u(L) = 0$. There is not much point in including these in the vector $x$, which contains the values of the function at the points where it is unknown. A common practice is to choose $\{s_i,\ i = 1, \cdots, d\} \subset (0, L)$ and approximate the second derivative at $s_1$ by $u''(s_1) = \frac{1}{h^2}(u(s_2) - 2u(s_1) - u(0)) = \frac{1}{h^2}(u(s_2) - 2u(s_1))$ and similarly for $s_d$. This approximation justifies the first and the last row of the matrix $L$. Other boundary conditions can also be accommodated in this way, for example, non-homogeneous Dirichlet conditions or Neumann boundary conditions. In the case of inhomogeneous Dirichlet boundary conditions, the boundary conditions are included in the formulation as an RHS, i.e., in $\mathscr{C}^c = \{V > \psi\}$, the solution of the VI solves $(L + k^2 I)x = b$, where $b$ is the vector that includes the boundary conditions.

Using the same discretization approach, we may also approximate the obstacle $\psi$ in terms of the finite dimensional approximations $\psi = (\psi_1, \cdots, \psi_d)$, where $\psi_i = \psi(s_i)$, and similarly for the function $f$ on the RHS of the VI. Then, condition $u \geq \psi$ is replaced by the finite dimensional approximation $x \geq \psi$, meaning that $x_i \geq \psi_i$ for every $i = 1, \cdots, d$. Like this we may replace the convex set $K := \{u \geq \psi\} \subset W_0^{1,2}(D)$ by a finite dimensional approximation $C = \{z \geq \psi\} \subset \mathbb{R}^d$. This is clearly a closed convex set, whose corresponding projection operator can be obtained in closed form as in (see Example 2.17)

$$\Pi_C(x) = (x - \psi)^+ + \psi = \max(x, \psi). \tag{3.105}$$

Using all the above we can obtain a finite approximation of the VI introduced in Proposition 1.25 in terms of

$$x^* \in C\ :\ \langle F(x^*), x - x^* \rangle \geq 0, \quad \forall x \in C = \{z \geq \psi\} \subset \mathbb{R}^d, \tag{3.106}$$

where $F(x) = (L + k^2 I)x - f$, with the vector $f$ containing the contribution $b$ of the boundary terms.

Any of the proposed schemes of this chapter can be used for the numerical treatment of the resulting finite dimensional VI. Moreover, the fact that the projection operator $\Pi_C$ is given explicitly in (3.105) makes the use of projection like or operator splitting methods attractive.

**Example 3.40 (The Projection Method)** The projection method (Algorithm 3.1) assumes the simple form

$$x_{n+1} = \max(x_n - \gamma((L + k^2 I)x_n - f), \psi).$$

This iterative scheme converges for a suitable choice of $\gamma > 0$. ◁

**Example 3.41 (Implicit Projection Method)** The implicit projection method (Algorithm 3.20) assumes the form

$$x_{n+1} = \max(x_n - \gamma[(L + k^2 I)x_{n+1} - f], \psi).$$

One way to implement this is to solve at any step the linear system

$$w_{n+1} = x_n - \gamma[(L + k^2 I)w_{n+1} - f] \iff [(1 + \gamma k^2)I + \gamma L]w_{n+1} = x_n - f \tag{3.107}$$

and then set

$$x_{n+1} = \begin{cases} w_{n+1} & \text{if} \quad w_{n+1} > \psi, \\ \psi & \text{otherwise.} \end{cases}$$

The solution of the linear system (3.107) can be performed by any method of choice (direct or iterative). The usual practice is to use iterative methods, such as the successive overrelaxation (SOR) scheme.[13] Often, the SOR steps are merged with the projection step, i.e., rather than waiting for the SOR part to terminate and apply the projection, each SOR step is interchanged with a projection step. Such a variant is the well-known projected SOR (PSOR) scheme. The results of its application to the perpetual American put are shown in Fig. 3.1. ◁

**Example 3.42 (The Douglas-Rachford Scheme)** The Douglas-Rachford Splitting scheme (Algorithm 3.28) becomes

$$\begin{cases} w_{n+1} = \max(z_n, \psi) = (z_n - \psi)^+ + \psi, \\ x_{n+1} = (I + \gamma F)^{-1}(2w_{n+1} - z_n), \\ z_{n+1} = z_n + \lambda_n(x_{n+1} - w_{n+1}) \end{cases} \iff$$
$$\begin{cases} w_{n+1} = \max(z_n, \psi) = (z_n - \psi)^+ + \psi, \\ (I + \gamma(L + k^2 I))x_{n+1} - f = (2w_{n+1} - z_n), \\ z_{n+1} = z_n + \lambda_n(x_{n+1} - w_{n+1}), \end{cases} \tag{3.108}$$

with the second step performed using, e.g., an SOR scheme. ◁

Any other method presented in this chapter can be used. Moreover, similar discretization schemes can be implemented for other operators, involving non-constant coefficients. Finally, the parabolic problem can also be treated, by combining the above methodology with a time stepping scheme.

[13] Recall that the SOR scheme for the solution of a linear system $Mx = b$ is in terms of the iterative scheme $(D_M + \omega L_M)z_{k+1} = [(1 - \omega)D_M - \omega R_M]z_k + \omega b$, where $D_M$, $L_M$, and $R_M$ are the diagonal, lower, and upper diagonal part, respectively, of the matrix $M$ and $\omega \in (0, 2)$ is a relaxation parameter chosen to accelerate convergence.

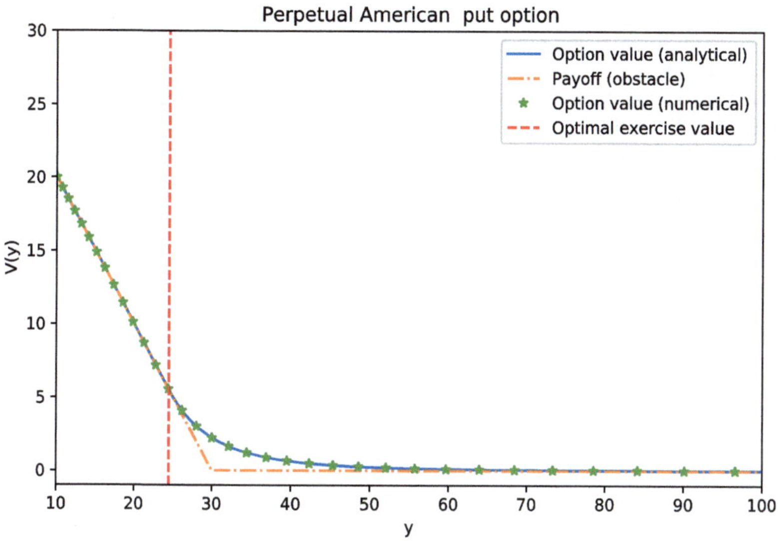


**Fig. 3.1** The analytical solution (solid line) versus the numerical approximation ($*$) for the perpetual American put option, using the PSOR method. The parameters of the model are chosen as $r = 0.05$, $\sigma = 0.15$

# References

Auslender A, Teboulle M (2000) Lagrangian duality and related multiplier methods for variational inequality problems. SIAM J Optim 10(4):1097–1115

Bauschke HH, Combettes PL (2017) Convex analysis and monotone operator theory in Hilbert spaces. CMS Books in Mathematics

Bermúdez A, Moreno C (1981) Duality methods for solving variational inequalities. Comput Math Appl 7(1):43–58

Ekeland I, Temam R (1999) Convex analysis and variational problems. SIAM

Glowinski R (2014) On alternating direction methods of multipliers: a historical perspective. Modeling, simulation and optimization for science and technology, pp 59–82

Goldstein A (1964) Convex programming in Hilbert space. Bull Am Math Soc 70:709–710

He B, Liao LZ (2002) Improvements of some projection methods for monotone nonlinear variational inequalities. J Optim Theory Appl 112:111–128

He B, Xu S, Yuan X (2023) Extensions of ADMM for separable convex optimization problems with linear equality or inequality constraints. In: Handbook of numerical analysis, vol 24. Elsevier, pp 511–557

Korpelevich GM (1976) The extragradient method for finding saddle points and other problems. Matecon 12:747–756

Kravvaritis DC, Yannacopoulos AN (2020) Variational methods in nonlinear analysis. De Gruyter

Kyritsi-Yiallourou ST, Papageorgiou NS (2009) Handbook of applied analysis, vol 160. Springer

Levitin ES, Polyak BT (1966) Constrained minimization methods. USSR Comput Math Math Phys 6(5):1–50

Lions PL, Mercier B (1979) Splitting algorithms for the sum of two nonlinear operators. SIAM J Numer Anal 16(6):964–979

Martinet B (1970) Brève communication. régularisation d'inéquations variationnelles par approximations successives. Revue française d'informatique et de recherche opérationnelle Série rouge 4(R3):154–158
Rockafellar RT (1976) Monotone operators and the proximal point algorithm. SIAM J Control Optim 14(5):877–898
Ryu EK, Yin W (2022) Large-scale convex optimization: algorithms & analyses via monotone operators. Cambridge University Press
Sion M (1958) On general minimax theorems. Pac J Math 8(1):171–176
Tseng P (2000) A modified forward-backward splitting method for maximal monotone mappings. SIAM J Control Optim 38(2):431–446

# Chapter 4
# Applications

*"It seems like when I get going, I just don't know how to stop"*

*John Coltrane*

*"Why don't you try taking the horn out of your mouth"*

*Miles Davis*

## 4.1 Introduction

In this chapter we show how the concepts and techniques introduced in this book can be extended and modified to deal with more advanced topics and models following selected recent developments in the research literature. Our selection of topics (which is short in the interest of brevity of the text) includes applications to VIs related to the pricing of American or real options in the presence of:

- Stochastic volatility (see Lamberton and Terenzi 2019)
- Multiple scale models (see Tsekrekos and Yannacopoulos 2016a,b)
- Model uncertainty (see Cheng and Riedel 2013)

In contrast to the previous chapters, here we do not provide detailed proofs for the quoted results but rather focus on indicating how these can be obtained by the appropriate extensions of the techniques developed in the previous three chapters and refer directly to the original papers for the proofs. To facilitate the reader in following the proofs, we keep as much as possible the original notation of the papers.

## 4.2 The American Put Option Under Stochastic Volatility

As mentioned in Sect. 1.2.3, the geometric Brownian motion (GBM) model (1.34) is too restrictive as a model for the underlying asset dynamics for models of American or real options. Phenomena such as volatility clustering, etc., which are typically



A. E. Tsekrekos, A. N. Yannacopoulos, *Variational Inequalities in Management Science and Finance*, EURO Advanced Tutorials on Operational Research,
https://doi.org/10.1007/978-3-031-74119-7_4

present in the data, require the extension of the GBM model in order to make it a more realistic representation of the observed evolution of the asset returns. A typical class of models that performs well in such tasks is the class of stochastic volatility models, which no longer assume that the volatility $\sigma$ is a constant, but rather a stochastic process driven by a (possibly unobservable) process $Y$. In this section we show how the theory of VIs developed in Chap. 2 can be extended to the pricing of American or real options for this class of asset models. The following comments are in order: The asset process now is two-dimensional, as the state of the system is fully determined by the pair $(s, y)$. The treatment of VIs in dimensions higher than 1 is not a problem, as the corresponding theory has already been provided in Chap. 2. The new concept that will be introduced here is allowing $(s, y)$ to take values on a possibly unbounded set. We will see that we can still formulate a variational formulation of the problem in this case, by a proper choice of Sobolev spaces.

Our exposition here follows Lamberton and Terenzi (2019). Assume the following dynamics, under the risk-neutral measure, for the underlying asset price $S$ and the volatility process $Y > 0$, which determines the stochastic volatility as $\sigma(Y) = \sqrt{Y}$,

$$\begin{aligned} dS_t &= (r - \delta)S_t dt + \sqrt{Y_t} S_t dB_t, \ S_0 = s > 0, \\ dY_t &= \kappa(\theta - Y_t)dt + \sigma\sqrt{Y_t} dW_t, \ Y_0 = y \geq 0, \end{aligned} \tag{4.1}$$

where $B$ and $W$ are two correlated Brownian motions, with correlation coefficient $\rho \in (-1, 1)$ (i.e., $d\langle B, W\rangle_t = \rho dt$), and $r \geq 0$, $\delta \geq 0$, $\kappa > 0$, $\theta > 0$ are parameters of the model. The dynamics of the volatility process are chosen to be of the Ornstein-Uhlenbeck type. This choice will lead to a stochastic volatility $\sigma(Y)$, displaying fluctuations around a mean state, a fact that seems compatible with stylized facts of real data.

In a similar fashion as in Sect. 1.2.3, it is more convenient to perform a change of variables similar to (1.40) and work with the process

$$X_t = \ln S_t - \bar{c}t, \ \text{with } \bar{c} = r - \delta - \frac{\rho\kappa\theta}{\sigma}. \tag{4.2}$$

In the new variables $(X, Y)$, as can be seen by a simple application of Itō's lemma, the model becomes

$$\begin{aligned} dX_t &= \left(\tfrac{\rho\kappa\theta}{\sigma} - \tfrac{Y_t}{2}\right) dt + \sqrt{Y_t} dB_t, \\ dY_t &= \kappa(\theta - Y_t)dt + \sigma\sqrt{Y_t} dW_t. \end{aligned} \tag{4.3}$$

We now consider the pricing of the standard American put on the asset $S$ for this model. In terms of the new variables $(X, Y)$ if exercised at $\mathfrak{s}$, this option has a payoff $\psi(\mathfrak{s}, X_{\mathfrak{s}}, Y_{\mathfrak{s}})$ where $\psi(t, x, y) = (K - e^{x+\bar{c}t})^+$, where $K$ is the exercise price. Hence, the price of the option at time $t$ (before exercise) if $X_t = x$ and $Y_t = y$ is given by $P(t, X_t, Y_t)$, with

$$P(t,x,y) = \sup_{\tau \in \mathscr{T}_{t,T}} \mathbb{E}\left[e^{-r(\tau-t)}\psi(\tau, X_\tau, Y_\tau) \mid X_t = x, Y_t = y\right], \tag{4.4}$$

where $\mathscr{T}_{t,T}$ is the set of all stopping times with values in $[t, T]$.

Using arguments similar in spirit to those led to the formulations (1.36) and (1.38) in Sect. 1.2.3, it can be shown that the discounted price function $\tilde{P}$ defined by the discounted price process $\tilde{P}(t, X_t, Y_t) = e^{-rt}P(t, X_t, Y_t)$ can be obtained as $\tilde{P}(t, X_t, Y_t) = \tilde{u}(t, X_t, Y_t)$, where $\tilde{u}$ is the solution of the following problem:

$$\frac{\partial \tilde{u}}{\partial t} + L\tilde{u} \leq 0, \quad \tilde{u} \geq e^{-rt}\psi, \quad \left(\frac{\partial \tilde{u}}{\partial t} + L\tilde{u}\right)\left(\tilde{u} - e^{-rt}\psi\right) = 0, \tag{4.5}$$

a.e. in $[0, T] \times D$, for $D := \mathbb{R} \times (0, \infty)$, supplemented with the final condition $\tilde{u}(T, x, y) = e^{-rT}\psi(T, x, y)$. In the above, $L$ is the operator

$$L = \frac{y}{2}\left(\frac{\partial^2}{\partial x^2} + 2\rho\sigma\frac{\partial^2}{\partial y \partial x} + \sigma^2\frac{\partial^2}{\partial y^2}\right) + \left(\frac{\rho\kappa\theta}{\sigma} - \frac{y}{2}\right)\frac{\partial}{\partial x} + \kappa(\theta - y)\frac{\partial}{\partial y}, \tag{4.6}$$

acting on functions $(x, y) \in D \mapsto u(x, y) \in \mathbb{R}$. Note that $L$ is the generator of the Itō process $(X_t, Y_t)$ defined in (4.3). Working in terms of the function $\bar{u}$ such that $\tilde{u}(t, x, y) = e^{-rt}\bar{u}(t, x, y)$ and the new time variable $\bar{t} := T - t$, signifying time to expiry, leads to an equivalent formulation of (4.5) as

$$\frac{\partial \bar{u}}{\partial \bar{t}} + A\bar{u} \geq 0, \quad \bar{u} \geq \psi, \quad \left(\frac{\partial \bar{u}}{\partial \bar{t}} + A\bar{u}\right)(\bar{u} - \psi) = 0, \tag{4.7}$$

a.e. in $[0, T] \times D$, for $D := \mathbb{R} \times (0, \infty)$, supplemented with the initial condition $\bar{u}(0, x, y) = \psi(0, x, y)$, where $A = -L + rI$. Upon dropping the bars for notational convenience, we see that (4.7) is of the form (1.38). Setting $(x_1, x_2) = (x, y)$, we observe that the operator $A$ is in the general form of the operators introduced in Definition 2.55 (see also Remark 2.57). However, we will stick to the notation $(x, y)$ (used in Lamberton and Terenzi (2019)) as it makes it easier to separate the observed variable $x$ (corresponding to the logarithm of the asset price) from the unobserved volatility variable $y$. As in Proposition 1.25, we can connect problem (4.7), formally, with a VI, and then using the results of Chap. 2, we can obtain a variational formulation of problem (4.7), leading to well-posedness and regularity results.

This programme requires the proper functional analytic setting for the problem, i.e., to clarify on which function space the operator $A$ acts. There are some technical subtleties that need to be addressed. These are related to the following issues: (a) The domain $D$ on which the function $u$ is defined is no longer bounded and (b) if we approach $y \to 0$, the operator $A$ loses the second order part, i.e., the ellipticity condition in Assumption 2.59 may not hold for such values of $y$ (this is often referred to as degeneracy of the operator). This is problematic, as it may lead to loss of the coercivity property, which was very important for our treatment of solvability and regularity of the VI.

To rectify this situation, we must choose a functional setting that remedies all the above. In Lamberton and Terenzi (2019), a weighted Sobolev space setting is introduced and designed so that the possible problematic behaviour of the solution of the VI for $|x| \to \infty$ and $y \to 0$ or $y \to \infty$ is "tamed." This is common practice in cases where unbounded domains or degenerate problems are to be encountered (see also Bensoussan and Lions (2011), Daskalopoulos and Feehan (2016), etc.).

Consider the measure

$$\mathfrak{m}(dx, dy) := \mathfrak{m}_{\gamma,\mu}(dx, dy) := y^{\beta-1} e^{-\gamma|x|-\mu y} dx dy, \tag{4.8}$$

with $\gamma > 0$, $\mu > 0$, and $\beta := \frac{2\kappa\theta}{\sigma^2}$, and consider functions such that their powers and their derivatives are integrable with respect to the density function related with (4.8). The density is exponentially weighted in $x$, controlling the behaviour of the solution for large $x$ (which may be unbounded), whereas the $y$ part of the density corresponds to the speed measure of the volatility process $Y$, related to the distribution of the random variables $Y_t$ for large enough $t$. The power law dependence of $\mathfrak{m}$ with respect to $y$ may absorb pathological behaviour of the solution for $y \to 0$, resulting from the fact that in this neighbourhood the diffusion coefficient of the $Y$ process vanishes (degeneracy), whereas the exponential part accounts for possible unbounded behaviour for large $y$. This is achieved by requiring that the class of possible functions is such that $\int_D |u|^2 d\mathfrak{m}(dx, dy) < \infty$. The class of such functions is often referred to as a weighted Lebesgue space. In fact, the class of such functions comprises a Hilbert space

$$L^2(D, \mathfrak{m}) = \{u : D \to \mathbb{R} \ : \ \int_D |u|^2 d\mathfrak{m}(dx, dy) < \infty\},$$

$$\|u\|^2_{L^2(D,\mathfrak{m})} := \int_D |u|^2 d\mathfrak{m}(dx, dy), \quad \langle u, v\rangle_{L^2(D,\mathfrak{m})} := \int_D uv d\mathfrak{m}(dx, dy). \tag{4.9}$$

Moreover, assuming the (weak) derivatives of $u$ belong to the above weighted Hilbert space, we can define the notion of weighted Sobolev spaces. Such an example can be

$$H^1(D, \mathfrak{m}) := \{u \in L^2(D, \mathfrak{m}) \ : \ (1 + y)^{1/2} u, \ y^{1/2}|Du| \in L^2(D, \mathfrak{m})\}, \tag{4.10}$$

which is a Hilbert space when endowed with the norm

$$\|u\|^2_{H^1(D,\mathfrak{m})} := \|(1 + y)^{1/2} u\|^2_{L^2(D,\mathfrak{m})} + \|y^{1/2}|Du|\|^2_{L^2(D,\mathfrak{m})}, \tag{4.11}$$

arising from the inner product

$$\langle u, v\rangle_{H^1(D,\mathfrak{m})} := \int_D (yDu \cdot Dv + (1+y)uv)d\mathfrak{m}.$$

This Sobolev space is the weighted analogue of the space $W^{1,2}(D)$ we encountered in Chap. 2. Higher order weighted Sobolev spaces may be defined accordingly. These Sobolev spaces enjoy embedding properties, akin to the standard Sobolev spaces, for example, $H^1(D, \mathfrak{m}) \subset L^2(D, \mathfrak{m})$.

Using the above weighted Sobolev spaces, we can define a bilinear form $\mathbf{a}_r : H^1(D, \mathfrak{m}) \times H^1(D, \mathfrak{m}) \to \mathbb{R}$, which allows for a variational definition of the operator $A := -L + rI$, in the sense that $\langle Au, v\rangle = \mathbf{a}_r(u, v)$ for every $v \in H^1(D, \mathfrak{m})$. Similarly as in Sect. 2.6.2, the interpretation of the $\langle\cdot,\cdot\rangle$ is a duality pairing.

Following Lamberton and Terenzi (2019), we define the following bilinear form:

**Definition 4.1** For any $u, v \in H^1(D, \mathfrak{m})$, we define the bilinear form

$$\begin{aligned}\mathbf{a}_0(u, v) = \frac{1}{2}\int_D y\Big(&u_x v_x(x, y) + \rho\sigma u_x v_y(x, y) + \rho\sigma u_y v_x(x, y)\\ &+\sigma^2 u_y v_y(x, y)\Big)\, d\mathfrak{m}\\ &+ \int_D y\left(j_{\gamma,\mu}(x)u_x(x, y) + k_{\gamma,\mu}(x)u_y(x, y)\right)v(x, y)d\mathfrak{m},\end{aligned}$$

where

$$j_{\gamma,\mu}(x) = \tfrac{1}{2}(1 - \gamma sgn(x) - \mu\rho\sigma) \text{ and } k_{\gamma,\mu}(x) = \kappa - \tfrac{\gamma\rho\sigma}{2} sgn(x) - \tfrac{\mu\sigma^2}{2}. \tag{4.12}$$

A simple integration by parts argument shows that $\mathbf{a}_0$ is the bilinear form associated with the operator $-L$. By defining the bilinear form $\mathbf{a}_r(u, v) = \mathbf{a}_0(u, v) + r\langle u, v\rangle_{L^2(D,\mathfrak{m})}$, we obtain the variational definition of the operator $A$, in terms of $\langle Au, v\rangle = \mathbf{a}_r(u, v)$ for every $u, v \in H^1(D; \mathfrak{m})$.[1] In terms of $\mathbf{a}_r$, (4.7) can be restated in the weak formulation as in Definition 2.86, Eq. (2.84), Sect. 2.7.2, choosing $X = H^1(D; \mathfrak{m})$. Then, using the methodology developed in Sect. 2.7.3, suitably modified, the existence and required regularity of the solutions to the VI, so that they can represent solutions of the optimal stopping problem, related to the pricing of the American option can be obtained. The required modifications include

[1] This is interpreted in the sense that for every $u$ sufficiently smooth ($u \in H^2(D, \mathfrak{m})$), where $H^2(D, \mathfrak{m})$ is a weighted analogue of $W^{2,2}(D)$, and for every $v \in H^1(D, \mathfrak{m})$, we have $\langle Au, v\rangle_{L^2(D,\mathfrak{m})} = \mathbf{a}_r(u, v)$. This can be extended by density arguments for every $u \in H^1(D, \mathfrak{m})$, but at the cost of $Au$ being extended also to an element of a larger space, hence interpreting $\langle\cdot,\cdot\rangle$ as a duality pairing.

a more relaxed coercivity condition for $\mathbf{a}_0$ (see Proposition 3.4 in Lamberton and Terenzi (2019)) and a technical truncation argument (to account for the unbounded coefficients, which introduce some difficulties with the penalization procedure; see Section 3.2, op. cit.).

The main results of Lamberton and Terenzi (2019) are summarized as follows:

**Theorem 4.1 (Theorem 2.3, Lamberton and Terenzi (2019))** *Subject to suitable boundedness and regularity assumptions on the pay-off function*[2] $\psi$*, there exists a unique function* $u$ *such that* $u \in C([0,T]; L^2(D,\mathfrak{m})) \cap L^2([0,T]; H^1(D,\mathfrak{m}))$, $\frac{\partial u}{\partial t} \in L^2([0,T]; H^1(D,\mathfrak{m}))$*, and*

$$-\left\langle \frac{\partial u}{\partial t}, v-u \right\rangle_{L^2(D,\mathfrak{m})} + \mathbf{a}_0(u, v-u) \geq 0 \ \text{a.e. in } [0,T], \tag{4.13}$$

$$\forall\, v \in L^2([0,T]; H^1(D,\mathfrak{m})),\ v \geq \psi,$$

$$u \geq \psi \ \text{ a.e. in } [0,T]\times\mathbb{R}\times(0,\infty), \qquad u(T) = \psi(T).$$

Note that, following the original source, we provide the solvability result for the final value problem (i.e., for a form similar to (4.5)). This can be turned to an initial value problem in terms of the variable transformation $t \to T-t$. The conditions on $\psi$ are analogue to those required for Proposition 2.88, but in the weighted space context.

Finally, we can show a weighted space analogue of Proposition 2.89, which will allow us to conclude that the unique solution of the variational inequality in (4.14) is the solution of the optimal stopping problem, i.e., the (discounted) American put price.

**Theorem 4.2 (Theorem 2.4, Lamberton and Terenzi (2019))** *Under further regularity conditions on* $\psi$*, the solution* $u$ *of the variational inequality* (4.14) *associated with* $\psi$ *is continuous and coincides with the function* $u^*$ *defined by*

$$u^*(t,x,y) = \sup_{\tau \in \mathcal{T}_{t,T}} \mathbb{E}\left[\psi(\tau, X_\tau, Y_\tau) \mid X_t = x,\ Y_t = y\right].$$

## 4.3 The Value of Waiting to Invest (Under Stochastic Volatility with Fast Mean Reversion)

### 4.3.1 A General Framework

In a variety of interesting financial problems, the Markov process that models the asset value or the investment project depends upon a parameter, and the pricing

[2] See the work cited for the exact conditions.

often has to be performed for the whole parametric family of value processes. Let us denote this parameter by $\epsilon$, and use the notation $X_\epsilon := \{X_\epsilon(t),\ \epsilon > 0\}$ for the whole family of processes for any value of the parameter $\epsilon$, and $L_\epsilon$ for the corresponding generator. This leads to a whole family of variational inequalities

$$\max\{L_\epsilon V_\epsilon(x) - r V_\epsilon(x), U_\epsilon(x) - V_\epsilon(x)\} = 0, \quad \epsilon > 0, \tag{4.14}$$

where the assumption is made that the utility or pay-off function $U$ may also depend on the same parameter $\epsilon$. Note that upon taking a minus sign (4.14) assumes the form (1.43). For problems such as (4.14), an interesting question that is often encountered is the behaviour of the whole family of solutions at a relevant limit, e.g., as $\epsilon \to 0$.

A few concrete examples of the above general formulation could be (a) parametric models of additional stochastic factors (e.g., stochastic volatility, stochastic short rate, etc.) where $\epsilon$ plays the role of a particular parameter whose value $\epsilon = 0$ corresponds to some benchmark model (e.g., the constant volatility and constant short rate Black and Scholes (1973) model) or (b) multi-scale stochastic volatility models, as in Fouque et al. (2003a), that will also be treated in the example of this section. Note that if we work in terms of the operator $A_\epsilon = -L_\epsilon + rI$ as in Chap. 2, we can get (4.14) in the standard form that was treated there.

We can conveniently formulate the family of problems in (4.14) as follows: Define the set $K_\epsilon = \{V \in X : V(x) \geq U_\epsilon(x), x \text{ a.e. } \in \mathbb{R}^d\}$, where $X$ is a conveniently chosen function space (most usually a Sobolev space). For $U_\epsilon$ continuous, $K_\epsilon$ is a closed, convex subset of $X$. Consider any function $\phi \in Z$, where $Z$ is a set of smooth functions, dense in $X$ (e.g., $Z = C_0^\infty(D)$). We multiply the inequality by any such $\phi$ and integrate over $\mathbb{R}^d$, to obtain the so-called weak form which reads

$$\langle A_\epsilon V_\epsilon, \phi - V_\epsilon \rangle \geq 0, \quad \forall \phi \in K_\epsilon, \tag{4.15}$$

where $A_\epsilon = -L_\epsilon + rI$. By $\langle \cdot, \cdot \rangle$ we denote the duality pairing $\langle \phi, \psi \rangle = \int_D \phi(x)\psi(x)dx$ between $X$ and its dual $X'$. We may simplify notation by using the family of bilinear forms $\mathbf{a}_\epsilon : X \times X \to \mathbb{R}$, defined by $\mathbf{a}_\epsilon(\phi, \psi) := \langle A_\epsilon \phi, \psi \rangle$, to express inequality (4.15) as

$$\mathbf{a}_\epsilon(V_\epsilon, \phi - V_\epsilon) \geq 0, \quad \forall \phi \in K_\epsilon. \tag{4.16}$$

Recall (see Definition 2.58) that for any $\epsilon > 0$, we say that the bilinear form $\mathbf{a}_\epsilon$ is continuous on $X$ if there exists a constant $c$ such that $\mathbf{a}_\epsilon(\phi, \psi) \leq c\|\phi\|_X \|\psi\|_X$ and coercive on $X$ if there exists a constant $c_0 > 0$ such that $\alpha_\epsilon(\phi, \phi) \geq c_0 \|\phi\|_X^2$. These notions can be extended to the whole family.

All that is essentially a simple extension of the framework developed in Chap. 2, for the above parametric family of problems. Using the techniques given in Chap. 2, e.g., the Lions-Stampacchia theory (see Theorem 2.62) the well-posedness of the weak formulation in (4.15) it can be shown (see, e.g., Tsekrekos and Yannacopoulos (2016a)).

**Proposition 4.2** *Assume that for any* $\epsilon > 0$*, the family of bilinear forms* $\alpha_\epsilon : X \times X \to \mathbb{R}$ *is continuous and coercive. Then for any* $\epsilon > 0$*, there exists a unique solution of the variational inequality* (4.16).

### 4.3.2 Multiple Scale Problems

While the above proposition provides the existence of a solution, and hence, subject to extra regularity, of a stopping rule for every $\epsilon$ in a quite general and intuitive way, the actual calculation of the stopping rule and the value function may be impossible in closed form. Moreover, an interesting question (which is also related to the actual calculation of the solution and the stopping rule) is its behaviour as $\epsilon$ tends to a relevant limit, e.g., $\epsilon \to 0$. For example, within a volatility modelling context, this limit could correspond to the case of high-frequency volatility variations. The study of limiting cases for the parameter $\epsilon$ leads us to problems related to homogenization theory (see Cioranescu and Donato 1999). Asymptotic arguments have to be employed in order to provide closed or semi-closed-form solutions for the solution of the variational inequality as $\epsilon \to 0$.

As an indication, assume that the process $X_\epsilon(t)$ can be split into two components, which evolve in different timescales, a "slow" and a "fast" one. Let us denote by $S$ and $Y$ the slow and fast components of $X_\epsilon$, respectively. We will use the notation $\epsilon = \delta^2$ and assume that $X_\epsilon = (S, Y)$ is modelled by the Itō process

$$dS_t = b(S_t, Y_t)dt + f(S_t, Y_t)dW_t^S, \quad S_0 = s \tag{4.17}$$

$$dY_t = \delta^{-2}\varsigma(Y_t)dt + \delta^{-1}\xi(Y_t)dW_t^Y, \quad Y_0 = y, \tag{4.18}$$

where the correlation structure between $dW_t^S$ and $dW_t^Y$ is described by

$$\begin{bmatrix} W_t^S \\ W_t^Y \end{bmatrix} = \begin{bmatrix} 1 & 0 \\ \rho & \sqrt{1-\rho^2} \end{bmatrix} \mathbf{W}_t, \tag{4.19}$$

with $\mathbf{W}_t$ a standard two-dimensional Wiener process on a complete filtered probability space satisfying the usual conditions (e.g., see Karatzas and Shreve 1991) and $|\rho| < 1$ a constant correlation coefficient. We will also use the notation $x = (s, y)$.

In the above, $Y_t$ is a latent stochastic factor that affects the dynamics of the observed component $S_t$ through appropriately chosen functions $b$ and $f$. The small parameter $\delta$, which is related to $\epsilon$ by $\epsilon = \delta^2$, models the fact that the latent factor $Y$ follows a process, whose dynamics are on a "faster" timescale than the dynamics of $S$.

We will assume that the "fast" component $Y$ reaches a statistical equilibrium at a timescale which is shorter than the scale of evolution of $S$. Therefore, after the passage of an initial transition period, the $Y$ part of the process $X_\epsilon$ will have reached an equilibrium density, thus allowing a decoupling of the system of equations (4.17)–(4.18), by replacing $Y$ in (4.17) with a constant random variable distributed by the equilibrium density, hence leading to a single equation for $S$. This intuitive but simple argument can be turned into a rigorous mathematical approach, leading to a perturbative, semi-closed-form solution for the class of variational inequalities associated with Markov processes $X_\epsilon = (S, Y)$ of the above form.

For the Markov process $X_\epsilon$, the generator operator is of the form

$$L_\delta = \delta^{-2} L_0 + \delta^{-1} L_1 + L_2, \tag{4.20}$$

with

$$L_0 = \zeta(y)\frac{\partial}{\partial y} + \frac{1}{2}\xi^2(y)\frac{\partial^2}{\partial y^2},$$

$$L_1 = \rho\xi(y) f(s, y)\frac{\partial^2}{\partial s \partial y},$$

$$L_2 = \frac{1}{2} f^2(s, y)\frac{\partial^2}{\partial s^2} + b(s, y)\frac{\partial}{\partial s},$$

leading to an expansion for the operator $A_\delta$ as $A_\delta = \delta^{-2}A_0 + \delta^{-1}A_1 + A_2$, where $A_0 = L_0$, $A_1 = L_1$, and $A_2 = L_2 - rI$, with $I$ the identity operator. For simplicity, we assume no explicit dependence of the function $U : \mathbb{R}^d \to \mathbb{R}$ on $\delta$. We further assume that $\zeta$ and $\xi$ are chosen so that the only acceptable solution of $L_0\psi = 0$ is independent of $y$ (Liouville property).

Under the above assumptions, it can be shown that if the variational inequality

$$\max\{-A_\delta V_\delta(x), U(x) - V_\delta(x)\} = 0, \quad \delta > 0, \quad x = (s, y), \tag{4.21}$$

admits a solution of the form

$$V_\delta(x) = \sum_{n=0}^{\infty} \delta^n V_n(x), \tag{4.22}$$

and the stopping rule is defined in terms of the solution of the equation $U(x) = V^\delta(x)$. The solvability of (4.21) follows from Proposition 4.4 (and related regularity results in the spirit of Proposition 2.77, Sect. 2.6.6), but the expansion (4.22) is at this point an assumption as there is no a priori reason as to why the dependence of $V$ on $\delta$ admits an expansion in powers of $\delta$. However, the validity of the expansion can be shown using techniques such as in Friedman and Hu (1997), Bensoussan et al (2011), or Bardi et al (2010) for the case of viscosity solutions, etc.

To search for the solution, the following definitions will be needed:

- Let $\varphi$ be the solution of the differential equation $L_0^*\varphi = 0$, where $*$ denotes the adjoint operator $\langle L_0^* u, v\rangle = \langle u, L_0 v\rangle$ for every suitable $u, v$. The solution $\varphi$ corresponds to the invariant density for the fast component $Y$ (normalized).
- Let $\overline{A}_2$ be the "averaged" operator

$$\overline{A}_2 = -\frac{1}{2}\overline{f^2}(s)\frac{\partial^2}{\partial s^2} - \overline{b}(s)\frac{\partial}{\partial s} + rI, \tag{4.23}$$

where for any function $g(s, y)$, $\overline{g}(s) := \int g(s, y)\varphi(y)dy$, the averaging over the fast variable $y$ over the invariant density for the fast component $Y$.

Then the following proposition can be proven (see Tsekrekos and Yannacopoulos (2016a)).

**Proposition 4.3** *$V_0$ and $V_1$ depend on $s$ only.*

*(i) $V_0$ solves the "averaged" variational inequality*

$$\max[-\overline{A}_2 V_0, U - V_0] = 0. \tag{4.24}$$

*(ii) $V_1$ solves the "averaged" equation of the form $\overline{A}_2 V_1 = \Psi(s, V_0)$, with the function $\Psi$ determined in terms of the data of the problem and $\varphi$ as integrals over $y$ (not reproduced here for the sake of brevity).*

***Proof of Sketch*** The proof relies on formally substituting the ansatz (4.22) in the VI and matching coefficients of $\delta$. We concentrate on $\mathscr{C}^c$ in which the VI turns to an equation. This yields (a) to zeroth-order $A_0 V_0 = 0$, which leads to the conclusion that $V_0$ depends only on $s$, modulo a multiplicative constant possibly depending on $y$ that cannot be determined at this stage and (b) to next order (since $A_1 V_0 = 0$) to $A_0 V_1 = 0$ from which we conclude that $V_1$ depends on $s$ only as well (again modulo a multiplicative constant possibly depending on $y$). The next order, still in $\mathscr{C}^c$, yields $A_0 V_2 = -A_2 V_0 - A_1 V_1 = -A_2 V_0$. This is now a non-homogeneous equation, for $V_2$ that admits a solution as long as the RHS is properly selected. We should note here that the RHS may also depend on $y$, due to the free multiplicative constants of $V_0$ or $V_1$ which we were not able to determine in the previous two stages. The compatibility condition for solvability of the equation for $V_2$ is provided by the Fredholm alternative. Put simply, just take the inner product of $A_0 V_2 = -A_2 V_0$ by a function $\varphi$ such that $A_0^*\varphi = 0$. Using the definition of $\phi$ and the operator $A_0$, this yields the compatibility condition for the RHS $\langle A_2 V_0, \varphi\rangle = 0$ which is interpreted as $\bar{A}_2 V_0 = 0$. For details see Tsekrekos and Yannacopoulos (2016a). □

One may proceed to higher order, in which dependence on both $s$ and $y$ is expected.

### 4.3.3 Fully Worked Example: Optimal Stopping Under "fast" Mean-Reverting Stochastic Volatility

We apply the result in Proposition 4.3 to a modified version of the classic "*waiting-to-invest*" model in McDonald and Siegel (1986). In their model, there is an *option to delay* an irreversible investment, so as to undertake it at some future point in time where its expected net present value is maximized. Such *real options* to defer investment in the face of uncertainty and cost irreversibility are typically encountered in many contexts (see, inter alia, for applications in natural resources, real estate, and R&D investments Brennan and Schwartz (1985); Paddock et al. (1988); Grenadier (1995); Schwartz and Moon (2000)) where managerial flexibility over the timing of the investment can be exerted.

In this setting, consider a risk-neutral decision-maker contemplating a new project that can be launched instantaneously, at any point in time once a constant fixed cost $I_0 > 0$ is incurred, and suppose that the estimated value of the project at time $t$, denoted $S_t$, evolves according to the Itō process in equations (4.17)–(4.18), with

$$b(S_t, Y_t) = \eta(\kappa - S_t)S_t,\ f(S_t, Y_t) = f\,(Y_t)\,S_t,\ \zeta(Y_t) = (m - Y_t) \text{ and } \xi(Y_t) = \frac{\nu\sqrt{2}}{\delta}. \tag{4.25}$$

Thus, the latent "fast" component $Y_t$ follows an Ornstein-Uhlenbeck (OU) process that depends on the small parameter $\delta$, while the observable "slow" component $S_t$ evolves according to a geometric OU process, whose variance is affected by the latent factor via the function $f$.

For any initial state $(s, y)$, the maximized expected present value of the project is

$$V(s, y) = \sup_{\tau > 0} \mathbb{E}\left[e^{-r\tau}[S(\tau) - I_0]\right]. \tag{4.26}$$

For the particular problem, $\varphi(y) = \frac{1}{\sqrt{2\pi}\,\nu} e^{-\frac{(m-y)^2}{2\nu^2}}$, and the "averaged" operator in (4.23) is

$$\overline{A}_2 = -\frac{1}{2}\overline{f}^2 s^2 \frac{\partial^2}{\partial s^2} - \eta(\kappa - s)s\frac{\partial}{\partial s} + rI,$$

with constant $\overline{f}^2 = \int_{-\infty}^{+\infty} f^2(y)\varphi(y)dy$ which is easy to calculate either analytically or numerically in terms of quadrature.

An application of Proposition 4.3 leads to the conclusion that the leading-order term $V_0$ is the solution of the variational inequality

$$\max[\frac{1}{2}\overline{f}^2 s^2 \frac{\partial^2 V_0}{\partial s^2} + \eta(\kappa - s)s\frac{\partial V_0}{\partial s} - rV_0, I_0 - V_0] = 0,$$

which yields (given that $S$ equals zero is an absorbing state and $\lim_{s\to 0} V_0(s) = 0$) a solution in $\mathscr{C}^c$ that can be expressed in terms of the confluent hypergeometric function $H(\beta, \alpha, x)$ (see Abramowitz and Stegun 1972) as $V_0(s) = As^\beta H(\beta, \alpha, \gamma s)$, with $\alpha$, $\beta$, and $\gamma$ depending on the data of the problem and $\bar{f}^2$. The constant $A$ is determined jointly with the leading-order term of the stopping rule threshold $S_0^*$ by the following "value-matching" and "smooth-pasting" conditions:

$$\begin{aligned} V_0(S_0^*) &= S_0^* - I_0, \\ V_0'(S_0^*) &= 1. \end{aligned} \tag{4.27}$$

Finally, for this particular problem the function $\Psi$ in Proposition 4.3 can be explicitly calculated.

In order to demonstrate our main result in Proposition 4.3, assume that the latent "fast" component $Y_t$ affects the dynamics of the project value in (4.25) via $f(Y_t) = \exp(Y_t)$. For this particular choice of $f(y)$, one gets $\overline{f} = \exp(m + \nu^2)$. We fix the parameters at $I_0 = \kappa = 1$, $r = \eta = 0.08$, $\delta = 1/\sqrt{200}$, $\nu = 1/\sqrt{2}$, $m = \ln 0.1$, and $\rho = -0.3$, as in Fouque et al. (2003b) and McDonald and Siegel (1986). For these parameter values, the blue solid line in Fig. 4.1 plots the leading-order value term $V_0(s)$ for $s \in [0, S_0^*]$, $S_0^* = 1.38849$. The dashed blue lines plot the "corrected" value function $V(s) + \delta V_1(s)$ for different correlation coefficients. We can see that "fast" stochastic volatility, which is usually negatively correlated with investment value, decreases the value of the option to invest and postpones optimal investment timing, and the effect is more pronounced, the more negative the correlation coefficient. The red and green lines demonstrate the effect of increasing $m$, which leads to an increase of the "averaged" volatility $\overline{f}$, making the option to invest more valuable, ceteris paribus.

## 4.4 Optimal Switching (Under Stochastic Volatility with Fast Mean Reversion)

An important class of problems arising in Management Science and Finance is the so-called optimal switching problems, in which the objective is to find the optimal time to initiate or terminate a process, subject to uncertainty. The literature is rich in applications of such problems in different contexts, ranging from manufacturing and logistics (e.g., Benaroch et al 2012) and production management and capacity choice (see, e.g., Dixit 1989; Trigeorgis 1993; Pindyck 1988; McDonald and Siegel 1985) to transportation and shipping (Sødal et al. 2008; Kavussanos and Tsekrekos

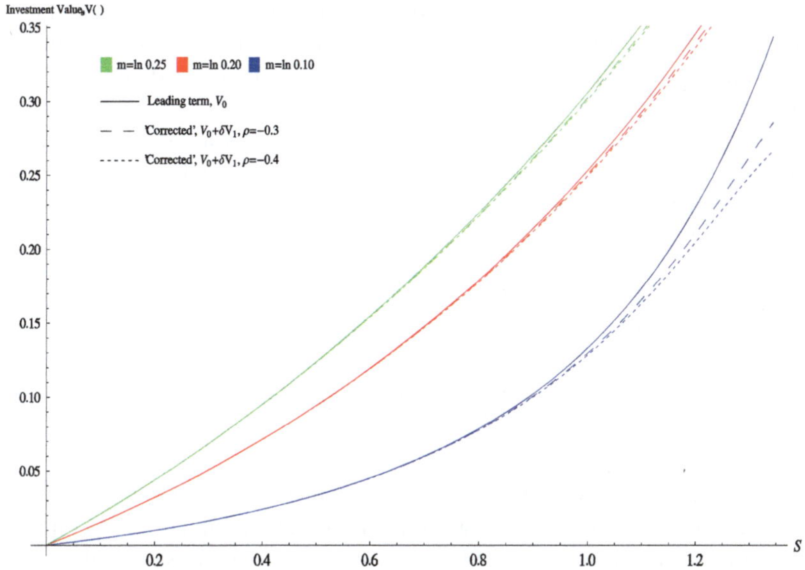


**Fig. 4.1** The figure plots (solid lines) the leading-order value term $V_0(s)$ and the "corrected" value $V(s) + \delta V_1(s)$ (dashed lines) for different parameter values. Lines in blue ($m = \ln 0.1$), red ($m = \ln 0.2$), and green ($m = \ln 0.25$) correspond to different long-run levels for the "fast" stochastic volatility latent factor. The rest of the parameters are $I_0 = \kappa = 1$, $r = \eta = 0.08$, $\delta = 1/\sqrt{200}$, $\nu = 1/\sqrt{2}$, and $f(y)$ is equal to $\exp(y)$

2011), to name a few. In this section we extend the investment ("entry-only") problem of the previous section, to the case where the investment project can be switched (initiated and terminated, at will). Our exposition follows Tsekrekos and Yannacopoulos (2016b).

Consider a risk-neutral decision-maker contemplating an investment project that can operate in two modes (active or idle) and which, depending on the mode, produces a stochastic flow pay-off $R_q(S)$ ($q = 0$ is idle, and $q = 1$ active), where $S$ is the investment project's underlying asset or price that can be modelled by a diffusion process. Transition from one mode to the other can take place instantaneously and for an unlimited number of times, but at constant fixed costs, each time a transition decision is made.

We assume that the model for the underlying $S$ of the project belongs to the general class of latent-factor stochastic volatility models as in (4.17)–(4.18). More specifically, we assume that the underlying $S$ and the stochastic volatility latent factor follow the processes in (4.17)–(4.18), with

$$b(S_t, Y_t) = \mu S_t,\ f(S_t, Y_t) = f(Y_t)\, S_t,\ \zeta(Y_t) = (m - Y_t) \text{ and } \xi(Y_t) = \tfrac{\nu\sqrt{2}}{\delta}, \tag{4.28}$$

where $f : I \subset \mathbb{R} \to \mathbb{R}$ ($I$ compact) is a smooth function.

The decisions to switch from one mode of operation to the other can be modelled by an adapted, finite variation, càglàd process $Q$ taking values in $\{0, 1\}$, with $Q_t = 0$ or 1 if at time $t$, the process is idle or active, respectively. Let $q$ denote the mode of operation at $t = 0$.

Let $K_0, K_1$ denote the fixed costs of *leaving* the idle or the active mode, respectively. For any switching decision rule $Q = (Q_t)_{t\geq 0}$ and initial state $(q, s, y)$, the present value of the project is

$$J_{(q,s,y)}(Q) = \mathbb{E}\left[\int_0^\infty e^{-rs}\left[R_1(S_s)\,Q_s + R_0(S_s)(1 - Q_s)\right]ds \right.$$
$$\left. - \sum_{s\geq 0} e^{-rs}\left[K_0(\Delta Q_s)^+ + K_1(\Delta Q_s)^-\right]\right], \qquad (4.29)$$

where $r$ is the (constant) risk-free rate of interest, and by $(x)^\pm$ we denote the positive/negative part of $x$. The flow functions $R_q$ are assumed to be concave and of the form $R_q(s) = a_q s + b_q + \mathfrak{R}_q(s)$, where functions $\mathfrak{R}_q$ satisfy the condition $\lim_{s\to 0}\frac{\mathfrak{R}_0(s)}{s} = \lim_{s\to+\infty}\frac{\mathfrak{R}_1(s)}{s} = 0$.

The objective is to maximize (4.29) over all possible switching rules $Q$ and obtain the value function

$$V_q^\delta(p, y) := \sup_Q J_{(q,p,y)}(Q), \qquad (4.30)$$

where the superscript $\delta$ is used to emphasize the dependence of the value function on the small parameter $\delta$.

The above general model provides a convenient and flexible framework for the treatment of a wide class of switching problems encountered in the operations research and management science literature. Examples that are treated as optimal switching problems in the literature range from supplier-switching options, which allow firms to switch suppliers and flexibly adjust their order quantity over time in the face of uncertainty in the underlying state process (e.g., the exchange rate in Kamrad and Siddique 2004), to manufacturing (Kogut and Kulatilaka 1994; Kazaz et al. 2005), to municipal waste management (Di Corato and Montinari 2014), and the optimal choice between short-lived and long-lived projects in the face of interest rate uncertainty (as in Dias and Shackleton 2011).

Using results from impulse control theory (see, e.g., Bensoussan and Lions 1984; Duckworth and Zervos 2001) and the perturbation method of Fouque et al. (2000), it can be shown that the value functions $V_q^\delta$ satisfy the quasi-variational inequality

$$\max\left\{L^\delta V_q^\delta(s,y)+R_q(s),\,V_{1-q}^\delta(s,y)-V_q^\delta(s,y)-K_q\right\}=0,\;\;q\in\{0,1\},\tag{4.31}$$

where the operator $L^\delta$ is the generator of the process $(S,Y)$, as in (4.20).

Assuming[3] that the variational inequality (4.31) admits a solution of the form

$$V_q^\delta(s,y)=\sum_{n=0}^{\infty}\delta^n V_{q,n}(s,y),\;\text{ and }\;S_q^\delta(y)=\sum_{n=0}^{\infty}\delta^n S_{q,n}(y),\tag{4.32}$$

the $V_{q,n}$ and $S_{q,n}$ are provided by the following proposition.

**Proposition 4.4** *For $n=0$, the value functions $V_{q,0}$ only depend on $s$, and in particular $V_{q,0}(s,y)=V_{q,0}(s)$ is the solution of the "averaged" equation*

$$\overline{L_2}V_{q,0}:=\frac{1}{2}\overline{f}^2s^2\frac{\partial^2 V_{q,0}}{\partial s^2}+\mu s\frac{\partial V_{q,0}}{\partial s}-rV_{q,0}=-R_q,\tag{4.33}$$

*subject to*

$$\lim_{s\to 0}V_{0,0}(s)=\frac{b_0}{r},\;\;\textit{and}\;\;\lim_{s\to+\infty}V_{1,0}(s)=\begin{cases}-\infty\;\textit{if}\;a_1\le 0,\\+\infty\;\textit{if}\;a_1>0,\end{cases}\tag{4.34}$$

*where* $\overline{f}^2=\lceil f^2(y)\rceil:=\int_{-\infty}^{+\infty}f^2(y)\frac{1}{\sqrt{2\pi}\nu}e^{-\frac{(m-y)^2}{2\nu^2}}dy$.

*The zeroth-order free boundary, $S_{q,0}$, is independent of $y$ and is provided by the solution of the algebraic system of four equations*

$$V_{1-q,0}\left(S_{q,0}\right)-V_{q,0}\left(S_{q,0}\right)=K_q,\tag{4.35}$$

$$\frac{\partial}{\partial s}\left[V_{1-q,0}(s)-V_{q,0}(s)\right]\Big|_{s=S_{q,0}}=0,\tag{4.36}$$

*where $V_{q,0}$ is the solution of* (4.33) *subject to* (4.34).

***Proof*** See Appendix A in Tsekrekos and Yannacopoulos (2016b). □

[3] An assumption that is intuitive but can be rigorously proved using techniques such as in Friedman and Hu (1997), Bensoussan et al (2011), or Bardi et al (2010).

Corrections to the first order result can also be calculated (see Tsekrekos and Yannacopoulos (2016b)).

### 4.4.1 Fully Worked Example: The Entry/Exit Problem Under Multi-scale Volatility

Consider the seminal, entry-exit/hysteresis model in Dixit (1989) where now

$$R_q(S_t) = \begin{cases} S_t - c & \text{for } q = 1, \\ 0 & \text{for } q = 0, \end{cases} \tag{4.37}$$

with $c$ a constant flow cost.

Using the result of Proposition 4.4, the following can be established:

**Proposition 4.5** *Up to the first order, the value function for the Dixit problem under the dynamics in* (4.17)–(4.18) *and* (4.28) *is* $V_q^\delta(s) = V_{q,0}(s) + \delta V_{q,1}(s)$, $q \in \{0, 1\}$, *where*

$$V_{q,0}(s) = (1-q)\, A s^\alpha + q\left(B s^\beta + \frac{s}{r-\mu} - \frac{c}{r}\right), \tag{4.38}$$

$$V_{q,1}(s) = \gamma \Gamma \left[(1-q)\, s^\alpha + q s^\beta\right]\left[(-1)^q D_q \ln s + C_q\right], \tag{4.39}$$

*with the constants* $\alpha, \beta, A, B, \Gamma, D_q, C_q$ *depending on the data of the problem and* $\bar{f}^2$, *leading to explicit characterization of the switching boundary subject to the solution of the algebraic system* (4.36)

Moreover, the first order corrections to the value functions and the switching boundary can be explicitly calculated (not reproduced here in the interest of brevity).

As an illustration of the general result, in Fig. 4.2 we illustrate it in the case where $f(y) = e^y$. We plot the optimal switching price thresholds with respect to the switching costs $K_q$ and the correlation coefficient $\rho$. Whether the effect of stochastic volatility on the thresholds is positive or negative depends on the correlation coefficient. For $\rho < 0$ which is the most usual case in financial variables (see, e.g., Fouque et al. 2000), the decision-maker of the investment project that can be switched from and to an idle/active mode will optimally decide to leave the current mode at lower thresholds: Lower levels of $S$ (than those in the constant volatility case with volatility set to $\overline{f}$) now lead to project activation, and the owner is now willing to incur more losses before activation is suspended on the down side.

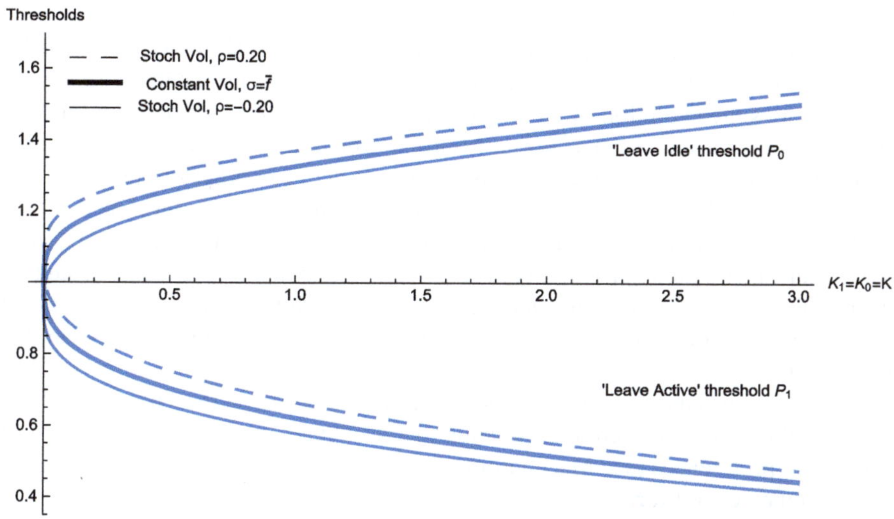


**Fig. 4.2** The figure plots $S_q$, $q \in \{0, 1\}$, i.e., the optimal switching thresholds under stochastic volatility with fast mean reversion and under the constant-volatility problem in Dixit (1989), with the constant $\sigma$ set equal to $\overline{f}$. The thresholds are plotted as a function of the switching costs $K_0$ and $K_1$, which are set to equal values, and for different values of the correlation coefficient $\rho$. The rest of the parameters are $r = 0.05$, $\mu = 0.04$, $m = \ln 0.1$, $\nu = 1/\sqrt{2}$, $c = 1$, and $\delta = 1/\sqrt{200}$

## 4.5 Value of Waiting to Invest Under Ambiguity

In many problems in economics and finance (e.g., timing of entry to a new market), it is difficult for the decision-maker to assess the probability distribution of future pay-offs. Situations of this type are often referred to as model uncertainty. This might be so because the market is quite new and few data on demand are available for reliable assessment of the future pay-off probability distribution. In other words, the decision-maker faces Knightian uncertainty (Knight 1921). Model uncertainty has become a rather hot topic in all fields of economics, finance, and management, leading to the development of interesting methodological tools and implications. Here we limit ourselves to the study of model uncertainty on optimal stopping and VIs. A theory of optimal decisions under Knightian uncertainty in discrete time has been developed by Riedel (2009) and in continuous time by Cheng and Riedel (2013), who we follow here.

We wish to study the optimal decision of a decision-maker who earns a reward $Y_\tau$ upon stopping at time $\tau$ where $Y = (Y_t)_{0 \le t \le T}$ is an adapted process describing the decision-maker's gain. If there was a single probabilistic model $P_0$ describing the evolution of the process $Y$, then, as in Sect. 1.2.3 the value of the decision/option at time $t$ and the optimal stopping time could be obtained by choosing the stopping time $\tau$ that maximizes $\mathbb{E}_{P_0}[Y_\tau \mid \mathscr{F}_t]$, where $\mathscr{F}_t$ is a $\sigma$-algebra containing information about the process up to time $t$. As we have seen, using dynamic programming, this criterion can lead us to a VI formulation for the determination of the value function of the problem and the stopping rule. However, in the realm of model uncertainty, we do not have the luxury of a single probabilistic model $P_0$ that describes the evolution of $X$; hence using $\mathbb{E}_{P_0}[Y_\tau \mid \mathcal{F}_t]$ as a decision making criterion does not make much sense. Multiple probability models may well serve as descriptions of $X$ (or various aspects of it), so a decision criterion reflecting that must be proposed, and based on that a theory of optimal stopping developed.

A standard practice is to replace the expectation $\mathbb{E}_{P_0}$ by the value $\mathcal{E}_t(Y_\tau)$ of a non-linear functional on probability space $\mathcal{E}_t$, chosen so that it takes into account all the possible probability models that may serve as descriptions of the process $X$ and the relevant information on the process up to time $t$. The question is what can this functional be.

Various proposals have been made in the literature as to the choice of $\mathcal{E}_t$. A very popular class of preferences (abiding to a nice axiomatic framework) is that of variational preferences (see Maccheroni et al. 2006a,b) which can be expressed (for $t = 0$) in the general form

$$\mathcal{E}_0(Y_\tau) = \inf_{P \in \mathcal{P}} E_P[Y_\tau + \zeta(P)], \tag{4.40}$$

for a penalty function $P \mapsto \zeta(P)$ defined on the set $\mathcal{P}$ of probability measures that may describe $Y$. Note that this can be rather general, for instance, we may choose $\mathcal{P} = \mathcal{P}(\mathbb{R})$ the set of probability measures on $\mathbb{R}$ if we model $X$ taking values in $\mathbb{R}$ directly or $\mathcal{P} = \mathcal{P}(\mathbb{R}^d)$ if we model $X$ in terms of a set of stochastic factors $Z$ taking values in $\mathbb{R}^d$, through some representation $Y = \Phi(Z)$ (as, e.g., in the GBM model in which we model the stochastic factor $Z = W$ and then we obtain the asset price $Y$ as a function of the factor process $W$, in terms of an exponential map). The economic intuition behind (4.40) is very natural; value your decision in terms of the worst-case scenario, under all possible models that may describe $X$, the contribution of each one moderated in terms of the penalty function $\zeta$, in the space of models. A close relative of variational preferences is convex risk measures (see, e.g., Föllmer and Schied (2011), Artzner et al (1999)) which are gaining popularity in theoretical and empirical studies in finance and insurance. For dynamic versions of (4.40), see Maccheroni et al. (2006b). The choice of penalty function $\zeta$ allows for a wide range of models of model uncertainty or ambiguity to be incorporated in (4.40). For example, if $\zeta$ is the convex indicator function of a set of priors $\mathcal{M} \subset \mathcal{P}$, then we recover the Gilboa-Schmeidler preferences (see Gilboa and Schmeidler 1989), $\mathcal{E}_0 = \inf_{P \in \mathcal{M}} \mathbb{E}_P(X_\tau)$, if $\zeta$ is chosen as the Kuhlback-Leibler divergence between $P$ and a reference measure $P_0$, then we recover the Hansen-Sargent multiplier preferences (see, e.g., Hansen and Sargent (2001), Hansen and Sargent (2008), Strzalecki (2011)), whereas if $\zeta$ is chosen in terms of the Fréchet function for a proper metrization of the space of probability measures (e.g., the Wasserstein metric), we obtain the Fréchet mean or barycentric preferences (see Petracou et al. (2022), Papayiannis and Yannacopoulos (2018)).

A common choice of dynamic non-linear functional $\mathcal{E}_t$ for the evaluation of an $\mathcal{F}_\tau$ measurable random variable $Y_\tau$ at time $t < \tau$ is the following. Assume that the possible probability measures $P$ that may describe the process $Y$ (and hence $Y_\tau$) are parameterized in terms of a set of measurable processes $\upsilon = (\upsilon_t)_{0 \leq t \leq T}$. This is not a bad assumption, at least in financial applications, where we often have to work under the risk-neutral measure, a task that is effected in terms of the Girsanov transformation, which parameterizes the set $\mathcal{P}$ in terms of an exponential transformation involving a stochastic process $\upsilon$, related to the change of drift in the evolution of $X$, when viewed under the alternative model (see, e.g., Oksendal (2013)). In this context we may denote each alternative model $P \in \mathcal{P}$ by $P(\upsilon)$, where $\upsilon \in \mathbb{D}$ is an adapted process, often called the information drift (taking values in an appropriate set $\mathbb{D}$), characterizing the measure, and express the non-linear functional $\mathcal{E}_t$ as

$$\mathcal{E}_t(Y_\tau) = \inf_{\upsilon \in \mathbb{D}} E_{P(\upsilon)}[X_\tau \mid \mathcal{F}_t + a_{t,\tau}(\upsilon)], \text{ where}$$

$$a_{t,\tau}(\upsilon) = \mathbb{E}_{P(\upsilon)}[\int_t^\tau f(s, \upsilon_s) ds \mid \mathcal{F}_t]. \tag{4.41}$$

The penalty function is therefore expressed in terms of the "rate" function $f$, depending on the information drift. Such a choice of the penalty function guarantees dynamic consistency for the evaluation functional $\mathcal{E}_t$. Moreover, there is an interesting connection of the representation (4.41) with the theory of backward stochastic differential equations (BSDEs), as the process $\mathcal{E} = (\mathcal{E}_t)_{0 \leq t \leq T}$ can be characterized as the solution of a BSDE (for more details, see Chen and Epstein (2002), Cheng and Riedel (2013), and Quenez and Sulem (2013)).

In terms of $\mathcal{E}$, the optimal stopping problem under model uncertainty (ambiguity) can now be expressed as finding the stopping time $\tau$ that gives the maximum value of the non-linear expectation $\mathcal{E}$ for the pay-off $Y$, when exercised. Following Cheng and Riedel (2013), we may define the value at time $t$, $V_t$ of the possibility to exercise and thus receive pay-off $Y_\tau$ at the stopping time $\tau \geq t$, as

$$V_t = \operatorname*{ess\,sup}_{\tau \geq t} \mathcal{E}_t(Y_\tau).$$

A probabilistic characterization of the process $V = (V_t)$ similar to the Snell envelope characterization in standard optimal stopping, leading to the characterization of the optimal stopping time $\tau^* = \inf\{t \geq 0 : V_t = X_t\}$, is provided in Theorem 3.1 in Cheng and Riedel (2013). Moreover, under certain conditions, the optimal stopping problem can be expressed in a saddle point formulation, maximizing over stopping times and minimizing over probability measures (models), which resembles a Nash equilibrium for a game between the decision-maker and a fictitious agent (nature) that chooses the probability models for $Y$. Under conditions, there exists a probability measure $P^*$ such that $V_t = \operatorname{ess\,sup}_{\tau \geq t} \mathbb{E}_{P*}[Y_\tau \mid \mathcal{F}_t]$ (Theorem 3.2, op. cit.).

What we are interested in here, and will focus on, is the connection of the optimal stopping problem under model uncertainty with VIs. For making this connection, we assume that the pay-off process $Y = (Y_t)_{0\leq t\leq T}$ can be expressed as $Y_t = \psi(t, X_t)$ for some function $t$, where $X = (X_t)_{0\leq t\leq T}$ is an Itō process, with values in $\mathbb{R}^N$, of the form

$$dX_s = b(X_s)dt + \sigma(X_s)dW_s, \quad X_t = x, \quad s > t, \tag{4.42}$$

where $W = (W_s)_{0\leq s\leq T}$ is an $m$-dimensional Wiener process, $b : \mathbb{R}^N \to \mathbb{R}^N$ and $\sigma : \mathbb{R}^N \to \mathbb{R}^{N\times m}$ are suitable drift and volatility functions. A simple example of such a case can be the basket option model in Sect. 1.2.3, Eq. (1.45), and the pay-off function for a put, e.g., $\psi(t, x_1, \cdots, x_N) = e^{-rt}\max(K - \sum_{i=1}^N w_i e^{x_i}, 0)$. Under standard conditions on $b$ and $\sigma$ (Lipschitz continuity and at most linear growth for $\sigma$), the process $X$ defined by (4.42) is a (strong) Markov process, with generator

$$L = \sum_{i=1}^N b_i(x)\frac{\partial}{\partial x_i} + \frac{1}{2}\sum_{i,j=1}^N \alpha_{ij}(x)\frac{\partial^2}{\partial x_i \partial x_j}, \tag{4.43}$$

where $\mathscr{A}(x) = (a_{ij}(x))_{i=1,\cdots,N}^{j=1,\cdots,N} = \sigma(x)\sigma^T(x)$. For such cases we can express the value process $V = (V_t)_{0\leq t\leq T}$ in terms of a function $v : [0, T] \times \mathbb{R}^N \to \mathbb{R}$ (denoted the same) such that $V_t = v(t, x)$. This is the analogue of the value function in the standard optimal stopping problem and can be determined as the solution of a suitable VI. In particular, the value function $v$ of the optimal stopping problem solves a Hamilton-Bellman-Jacobi equation of the type

$$\max_{(t,x)\in[0,T]\times\mathbb{R}^N} \{f(t, x) - v(t, x), v_t(t, x) + Lv(t, x)\} = 0,$$

where $L$ is the operator in (4.43).

The following verification theorem (Cheng and Riedel 2013) provides the extension of the HJB equation to the case of ambiguity.

**Theorem 4.6** *Assume that $\psi : [0, T] \times \mathbb{R}^N \to R$ be a continuous function that satisfies a polynomial growth condition in $x$, and define the function $g : [0, T] \times \mathbb{R}^N \to \mathbb{R}^N$ in terms of the Fenchel transform*

$$f(t, v) = \sup_{z\in\mathbb{R}^d} (g(t, z) - \langle z, v\rangle_{\mathbb{R}^N}).$$

*Let* $v : [0, T] \times \mathbb{R}^N \to R$ *be a* $W^{1,2}$ *function (satisfying along with its first derivatives a polynomial growth condition in* $x$*) that solves the non-linear Hamilton-Bellman-Jacobi equation*

$$\max_{(t,x)\in[0,T]\times\mathbb{R}^N} \left\{ f(t,x) - v(t,x), v_t(t,x) + Lv(t,x) + g(t, \sigma(x)^T Dv(t,x)) \right\} = 0,$$

$$v(T,x) = f(T,x). \tag{4.44}$$

*Then* $v$ *is the value function of our optimal stopping problem, i.e.,* $V_t = v(t, X_t)$.

The above verification theorem can be obtained also within the framework of viscosity solutions (see, e.g., Dumitrescu et al (2015), where it has also been extended for the case of jump diffusion processes).

## References


Abramowitz M, Stegun IA (1972) Handbook of mathematical functions. Dover Publications, New York

Artzner P, Delbaen F, Eber JM, Heath D (1999) Coherent measures of risk. Math Finance 9(3):203–228

Bardi M, Cesaroni A, Manca L (2010) Convergence by viscosity methods in multiscale financial models with stochastic volatility. SIAM J Financ Math 1(1):230–265

Benaroch M, Webster S, Kazaz B (2012) Impact of sourcing flexibility on the outsourcing of services under demand uncertainty. Eur J Oper Res 219(2):272–283

Bensoussan A, Lions JL (1984) Impulse control and quasi–variational inequalities. Gauthier–Villars, Paris

Bensoussan A, Lions JL (2011) Applications of variational inequalities in stochastic control. Elsevier

Bensoussan A, Lions JL, Papanicolaou G (2011) Asymptotic analysis for periodic structures, vol 374. American Mathematical Soc.

Black F, Scholes M (1973) The pricing of options and corporate liabilities. J Polit Econ 81(3):637–654

Brennan MJ, Schwartz ES (1985) Evaluating natural resource investments. J Bus 58(2):135–157

Chen Z, Epstein L (2002) Ambiguity, risk, and asset returns in continuous time. Econometrica 70(4):1403–1443

Cheng X, Riedel F (2013) Optimal stopping under ambiguity in continuous time. Math Financ Econ 7:29–68

Cioranescu D, Donato P (1999) An introduction to homogenization. Oxford lecture series in mathematics and its applications. The Clarendon Press/Oxford University Press, New York

Daskalopoulos P, Feehan PM (2016) C1, 1 regularity for degenerate elliptic obstacle problems. J Differ Equ 260(6):5043–5074

Di Corato L, Montinari N (2014) Flexible waste management under uncertainty. Eur J Oper Res 234(1):174–185

Dias JC, Shackleton MB (2011) Hysteresis effects under CIR interest rates. Eur J Oper Res 211(3):594–600

Dixit A (1989) Entry and exit decisions under uncertainty. J Polit Econ 97(3):620–638

Duckworth JK, Zervos M (2001) A model for investment decisions with switching costs. Ann Appl Probab 11(1):239–260

Dumitrescu R, Quenez MC, Sulem A (2015) Optimal stopping for dynamic risk measures with jumps and obstacle problems. J Optim Theory Appl 167:219–242

Föllmer H, Schied A (2011) Stochastic finance: an introduction in discrete time. Walter de Gruyter

Fouque JP, Papanicolaou G, Sircar KR (2000) Derivatives in financial markets with stochastic volatility. Cambridge University Press

Fouque JP, Papanicolaou G, Sircar R, Sølna K (2003a) Multi-scale stochastic volatility asymptotics. Multi–scale Model Simulat 2(1):22–42

Fouque JP, Papanicolaou G, Sircar R, Sølna K (2003b) Singular perturbation in option pricing. SIAM J Appl Math 63(5):1648–1665

Friedman A, Hu B (1997) A non-stationary multi-scale oscillating free boundary for the Laplace and heat equations. J Differ Equ 137(1):119–165

Gilboa I, Schmeidler D (1989) Maxmin expected utility with non-unique priors. J Math Econ 18:141–153

Grenadier SR (1995) Valuing lease contracts: A real options approach. J Financ Econ 38:297–331

Hansen LP, Sargent TJ (2001) Robust control and model uncertainty. Am Econ Rev 91(2):60–66

Hansen LP, Sargent TJ (2008) Robustness. Princeton University Press

Kamrad B, Siddique A (2004) Supply contracts, profit sharing, switching, and reaction options. Manag Sci 50(1):64–82

Karatzas I, Shreve SE (1991) Brownian motion and stochastic calculus. Springer, Berlin, Heidelberg, New York

Kavussanos M, Tsekrekos AE (2011) The option to change the flag of a vessel. International handbook of maritime economics, pp 47–62

Kazaz B, Dada M, Moskowitz H (2005) Global production planning under exchange-rate uncertainty. Manag Sci 51(7):1101–1119

Knight FH (1921) Risk, uncertainty and profit. Houghton Mifflin

Kogut B, Kulatilaka N (1994) Operating flexibility, global manufacturing, and the option value of a multinational network. Manag Sci 40(1):123–139

Lamberton D, Terenzi G (2019) Variational formulation of American option prices in the Heston model. SIAM J Financ Math 10(1):261–308

Maccheroni F, Marinacci M, Rustichini A (2006a) Ambiguity aversion, robustness, and the variational representation of preferences. Econometrica 74(6):1447–1498

Maccheroni F, Marinacci M, Rustichini A (2006b) Dynamic variational preferences. J Econ Theory 128(1):4–44

McDonald R, Siegel D (1986) The value of waiting to invest. Quar J Econ 101(4):707–727

McDonald RL, Siegel DR (1985) Investment and the valuation of firms when there is an option to shut down. Int Econ Rev, 331–349

Oksendal B (2013) Stochastic differential equations: an introduction with applications. Springer Science & Business Media

Paddock JL, Siegel DR, Smith JL (1988) Option valuation of claims on physical assets: The case of offshore petroleum leases. Quar J Econ 103(3):479–508

Papayiannis GI, Yannacopoulos AN (2018) Convex risk measures for the aggregation of multiple information sources and applications in insurance. Scand Actuar J 2018(9):792–822

Petracou EV, Xepapadeas A, Yannacopoulos AN (2022) Decision making under model uncertainty: Frechet-Wasserstein mean preferences. Manag Sci 68(2):1195–1211

Pindyck RS (1988) Irreversible investment, capacity choice and the value of the firm. Am Econ Rev 78(5):969–985

Quenez MC, Sulem A (2013) BSDEs with jumps, optimization and applications to dynamic risk measures. Stochast Process Appl 123(8):3328–3357

Riedel F (2009) Optimal stopping with multiple priors. Econometrica 77(3):857–908

Schwartz ES, Moon M (2000) Evaluating research and development investments. In: Brennan MJ, Trigeorgis L (eds) Project flexibility, agency, and competition : New developments in the theory and application of real options. Oxford University Press, New York, pp 85–106

Sødal S, Koekebakker S, Aadland R (2008) Market switching in shipping? A real option model applied to the valuation of combination carriers. Rev Financ Econ 17(3):183–203

Strzalecki T (2011) Axiomatic foundations of multiplier preferences. Econometrica 79(1):47–73

Trigeorgis L (1993) Real options and interactions with financial flexibility. Financ Manag, 202–224

Tsekrekos AE, Yannacopoulos AN (2016a) Approximation of optimal stopping problems and variational inequalities involving multiple scales in economics and finance. In: Pinto A, Accinelli Gamba E, Yannacopoulos A, Hervés-Beloso C (eds) Trends in mathematical economics. Springer, pp 317–329

Tsekrekos AE, Yannacopoulos AN (2016b) Optimal switching decisions under stochastic volatility with fast mean reversion. Eur J Oper Res 251(1):148–157

# Index





A. E. Tsekrekos, A. N. Yannacopoulos, *Variational Inequalities in Management Science and Finance*, EURO Advanced Tutorials on Operational Research,
https://doi.org/10.1007/978-3-031-74119-7

MIX
Papier aus verantwortungsvollen Quellen
Paper from responsible sources
FSC® C105338

If you have any concerns about our products,
you can contact us on
**ProductSafety@springernature.com**

In case Publisher is established outside the EU,
the EU authorized representative is:
**Springer Nature Customer Service Center GmbH**
**Europaplatz 3, 69115 Heidelberg, Germany**

Printed by Libri Plureos GmbH
in Hamburg, Germany